RESTORING CLAUSEWITZ

RESTORING CLAUSEWITZ

A Critical Companion to *On War*

J. Furman Daniel III

Rapid Communications in Conflict and Security Series
General Editor: Thomas G. Mahnken
Founding Editor: Geoffrey R. H. Burn

CAMBRIA PRESS

Amherst, New York

Requests for permission should be directed to
permissions@cambriapress.com, or mailed to
Cambria Press, University Corporate Centre,100 Corporate
Parkway, Suite 128, Amherst, New York 14226, USA.

Library of Congress Cataloging-in-Publication Data

Names: Daniel, J. Furman, III author

Title: Restoring Clausewitz : a critical
companion to "On war" / J. Furman Daniel, III.

Other titles: Critical companion to "On war"
Description: Amherst, New York : Cambria Press, [2025] |
Includes bibliographical references and index. |
Summary: "Restoring Clausewitz: A Critical Companion to On War
places Carl von Clausewitz's seminal treatise in historical context and
clarifies its core concepts, including friction, the trinity, political purpose,
and military genius. Addressing widespread misinterpretations and the
complexity of Clausewitz's dialectical method, this study offers a clear
guide to On War for students, military professionals, and general readers
interested in war, strategy, and military theory"-- Provided by publisher.

Identifiers: LCCN 2025018699 (print) | LCCN 2025018700 (ebook) |
ISBN 9781638573364 library binding | ISBN 9781638573616 paperback |
ISBN 9781638573623 epub | ISBN 9781638573630 pdf

Subjects: LCSH: Clausewitz, Carl von, 1780-1831. Vom Kriege |
Clausewitz, Carl von, 1780-1831 | Military art and science | War

Classification: LCC U102.C6643 D36 2025 (print) | LCC U102.C6643 (ebook)

LC record available at https://lccn.loc.gov/2025018699

LC ebook record available at https://lccn.loc.gov/2025018700

To those who made me laugh, I needed it

TABLE OF CONTENTS

ACKNOWLEDGEMENTS

It is not an exaggeration that this book would have been impossible without laughter. Laughter is more than an emotion. It is a tool, an escape, a release, a weapon, and so much more. While writing this book, there were many times that I worried that I was becoming a disgruntled war college professor, just like Carl von Clausewitz. Fortunately, I was surrounded by an outstanding collection of friends who were willing to smile and share moments of kindness and vulnerability. Indeed, laughter is much more than an emotion.

Special thanks to Dannielle Andrews, David Armstrong, Dave Arnold, Natalie Baker, Geoffrey Burn, Sam Cook, Sean Cumming, Mitchell Dallas, Jon "Bubba" Ehret, Elliott Fullmer, Todd Glasser, Gator Greenwill, Joe Holwell, Mike Jones, "Godfather" Bob McKenzie, Jay M. Parker, Dale Perez, Brian A. Smith, Bob Watts, and Michelle Wright. In countless ways, your intelligence, kindness, support, erudition, time, and humor are embedded in the pages of this book.

Finally, I owe a special debt to my family. Each of you has enriched my life and made this book possible. Christina, Claudia, Lawson, and Mom, I love you more than I can ever express. Words fail me.

RESTORING CLAUSEWITZ

LINCOLN AND THE
PRUSSIAN: A FABLE

President Abraham Lincoln needed a winning strategy. Despite the North's superior numbers, supplies, soldiers, and cause, a series of humiliating defeats had led many to question its ability to win the Civil War. By January 1862, the war effort had stalled, and the American experiment in democracy seemed on the verge of collapse. Even the typically stoic Lincoln implored his chief quartermaster, General Montgomery Meigs:

> General, what shall I do? The people are impatient; Chase has no money and tells me he can raise no more. The General of the Army has typhoid fever. The bottom is out of the tub...What shall I do?[1]

Despite so many material advantages, the Northern cause appeared to be derailed by poor strategy, uninspired leadership, and misfortune. Though not a military professional, Lincoln understood that he needed a new strategy—and quickly.

According to legend, the self-made Lincoln returned to the method that had long served him well since his hardscrabble boyhood in Kentucky and Illinois. He began to read and apply his common sense to better himself. Just as he had once studied the Bible by firelight, he now turned to tactical manuals and strategic treatises, eventually making himself a self-taught expert on military affairs. Among these texts, none was more influential—so the story goes—than *On War* by the Prussian General Carl von Clausewitz. Like a divine revelation, this weighty text offered the president insight into a wide range of issues, especially the need for clear political objectives and war aims. Armed with Clausewitz's wisdom, Lincoln tapped into his innate genius and quickly emerged as an effective wartime president. In this telling, the fate of the world hung by a slender thread—and history might have unfolded quite differently had the great president and the great strategist not, in some sense, collaborated to secure Union victory. According to this myth, Clausewitz played a critical—if unacknowledged—role in winning the American Civil War.

Unfortunately, much of this story is a compelling fiction with little basis in fact. It is true that the Union had significant material advantages and was initially hampered by poor strategy. It is also true that Lincoln was an unusually active wartime president who frequently read about military affairs and thought deeply about strategic matters. However, there is no evidence that Lincoln ever read *On War* or was even aware of the relatively obscure Prussian general. As will be discussed in greater detail, *On War* had not been translated into English at this time, Lincoln did not read German, the book's publication was extremely limited, and no known copies were present in North America at this time. Moreover, there is no record linking Lincoln or any of his senior advisors to the book. Simply put, the evidence does not support this appealing but unfounded narrative.[2] Yet despite these inconvenient facts, historians, biographers, novelists, and enthusiastic amateurs continue to perpetuate the view that the Northern cause benefited from the insights of the Prussian master.[3]

Though apocryphal, this tale underscores a central truth about Carl von Clausewitz: he is frequently—and often profoundly—misunderstood. If Clausewitz were merely a minor nineteenth-century German philosopher, such misunderstandings would matter little. Few today are troubled by the oversimplification or misinterpretation of the works of Friedrich Wilhelm Joseph Schelling or Gottlob Frege. Though important in their own time, these men are now largely forgotten. More importantly, few, if any, policymakers are attempting to apply their ideas to today's strategic challenges.

Clausewitz, by contrast, has neither been forgotten nor ignored. He wrote about the most consequential matters—war and peace—and his ideas continue to inform (and misinform) strategic decisions at the highest levels. If ideas shape strategic choices, and if those choices matter, then Clausewitz's work demands serious and proper engagement. Misunderstanding Clausewitz is uniquely dangerous because of both the nature of his subject and the ongoing influence of his work on real-world policy decisions.

This book seeks to restore Clausewitz by debunking the myths that surround his work, situating him in proper historical context, and making his ideas accessible and applicable to the challenges of our own time. In this case, the truth is indeed more compelling than the legend.[4]

Notes

1. Donald, *Lincoln*, 330.
2. Bassford, *Clausewitz in English*, 51–55.
3. For three prominent examples of the mythical connection between Lincoln and Clausewitz, see Stephenson, *Lincoln*; Sandburg, *Abraham Lincoln*; and Vidal, *Lincoln*. To be fair to Sandburg, he was a poet rather than a trained historian, and he likely approached these books as a form of American epic rather than as strictly factual biography. Nevertheless, his Pulitzer Prize–winning work remains one of the most influential and widely read accounts of Lincoln's life. See Hurt, "Sandburg's Lincoln Within History," 55–65. For a similarly problematic claim that the Union general George Meade was "well acquainted" with Clausewitz's and that his theories contributed to the Northern victory, see Brown, *Meade at Gettysburg*, 81, 105, 127, 321, 352–354, and 429–432.
4. One of the most interesting truths about Lincoln is that despite a lack of formal military training, he was able to become a superb wartime leader. See McPherson, *Tried by War*; Williams, *Lincoln Finds a General*; and Williams, *Lincoln and His Generals*.

WHY A NEW ENGAGEMENT WITH *ON WAR* IS NEEDED

The Mythical Name

The name Carl von Clausewitz holds near-mythic appeal for students of war. Its overtly Prussian and martial resonance seems to confer instant gravitas in strategic debates, serving as a shorthand for military authority and strategic genius. Indeed, his principal work, *On War*, has become a foundational text in professional military education worldwide. Engaging with its demanding content has become almost a rite of passage for aspiring military officers. Nearly every American military officer can quote (or misquote) a line about war as an act or force, a political instrument, or a product of friction. The Chinese military has also engaged seriously with Clausewitz's theories, as evidenced by the publication of more than thirty Chinese translations of *On War* in recent years.[1]

Yet reverence for Clausewitz extends well beyond military circles. His name and ideas are frequently invoked by figures as diverse as Vladimir Lenin, T. E. Lawrence, Henry Kissinger, Mao Zedong, Bob Dylan, Osama Bin Laden, and even fictional characters such as James Bond.[2] Since the end of World War II, Clausewitz has entered the mainstream, often serving as a proxy for military knowledge and authority. A partial list of recent cultural references includes works by Leo Tolstoy, C. S. Forester, C. S. Lewis, Ian Fleming, Timothy Findley, John G. Hemry, as well as major films such as *Lawrence of Arabia* (1962), *Cross of Iron* (1977), *Crimson Tide* (1995), *Lions for Lambs* (2007), and *Law Abiding Citizen* (2009).

Nobel Prize–winning writer and singer Bob Dylan lends further credence to the view of Clausewitz as a prophetic thinker deserving serious consideration. In his 2004 memoir, *Chronicles: Volume One*, Dylan claims that he discovered *On War* while perusing his friend's personal library during the period he was performing at the Gaslight Cafe in Greenwich Village. He admitted that from an early age he had "a morbid fascination with this stuff," that "Clausewitz in some ways is a prophet," and reading his works can make you "take your own thoughts a little less seriously."[3] By Dylan's own admission, Clausewitz's work prompted deeper reflection, even if *On War* did not share the pacifist themes found in Dylan's music.

Given this widespread influence and frequent citation, the legacy of Clausewitz would seem to be a settled matter. Indeed, by this measure, he appears to be the greatest writer on strategy who has ever existed. All hail von Clausewitz. The great philosopher of war. The final authority for strategy. The mythical name of Prussian military perfection revered in military, academic, and popular circles alike.[4] Indeed, it seems like there is little left to do but to stand back in awe of this great thinker who has so dominated our strategic vocabulary and mindset.[5]

Popular Acclaim Does Not Equal Proper Understanding

However, while it is relatively easy to demonstrate Clausewitz's popularity, it is far more difficult to prove that his ideas have been properly understood. Indeed, a popular quip about *On War* is that it is more often quoted than read. This is understandable. The text is both lengthy and dense. It is repetitive and, at times, internally contradictory. Much of the content focuses on now-obsolete forms of Napoleonic warfare that appear to hold limited relevance for contemporary readers, and its dialectical style resists straightforward interpretation. Compounding these issues is the fact that the book contains hundreds of eminently quotable lines, which offer aspiring strategists clever and seemingly authoritative excerpts. Unfortunately, the very quotability of *On War* has fostered distorted and reductive interpretations, reducing Clausewitz's complex ideas to oversimplified aphorisms and formulaic "best practices."

Similarly, many of the pop culture references cited earlier propagate highly inaccurate or misleading representations of Clausewitz's theories. The 1995 film *Crimson Tide*, starring Denzel Washington and Gene Hackman, provides a striking example. The film enjoys a dedicated following in the US Navy and is frequently cited in officer training programs, professional military education, and wardroom discussions.[6] This would be relatively harmless were the film merely a suspenseful narrative with a military theme and stylized, theatrical dialogue. Yet, the film inadvertently perpetuates an understanding of Clausewitz that is, at best, incomplete. Early in the film, the officers of the ballistic missile submarine *USS Alabama* discuss nuclear weapons and the meaning of the oft-cited phrase, "War is a continuation of politics by other means." This wardroom discussion serves as a catalyst for the film's central conflict and introduces the initial tension between the main characters. The Harvard-educated executive officer, Commander Hunter (Denzel Washington), challenges Captain Ramsey's (Gene Hackman) interpretation of Clausewitz as overly simplistic and outdated. Though ultimately unresolved and laced with humor, the exchange highlights a central contrast between the analytical, university-educated Hunter and the instinct-driven, traditionalist Ramsey.[7]

Although Commander Hunter advocates for a more nuanced reading of Clausewitz, it remains unclear whether he himself fully grasps the theorist's intended meaning. Indeed, Hunter ultimately concedes that "Clausewitz was trying to say" that "the sailor most likely to win the war is the one most willing to part company with the politicians and ignore everything except the destruction of the enemy."[8] This interpretation is deeply problematic on several levels. Clausewitz wrote virtually nothing about naval warfare and was unequivocal in asserting that war could never be entirely separated from politics. Although Clausewitz noted that war tended toward ever-increasing levels of violence, he presented this as a descriptive fact about the nature of war, not as a prescriptive guideline for policy. While screenwriters are not expected to produce scholarly interpretations of nineteenth-century military theory, the unintended consequence is significant: multiple generations of naval officers and millions of casual viewers have absorbed a caricatured version of Clausewitz, articulated by the film's charismatic and authoritative protagonist.

Unfortunately, the issue extends well beyond the limitations of Hollywood script writers. If that were only problem, this book would be unnecessary—just another case of academics lamenting the factual inaccuracies of works that enjoy far greater commercial success than their own. The more troubling reality is that although Clausewitz's ideas have deeply influenced military and political leaders, this influence has not consistently translated into sound strategy-making or decisive battlefield outcomes. Since Clausewitz's elevation to canonical status in American strategic thought, the United States has experienced a mixed—and often disappointing—record in its military engagements. With the possible exception of the 1991 Gulf War, the United States has struggled to achieve clear or enduring victories in its post–Cold War conflicts. Instead, it has been bogged down in a series of protracted and unpopular conflicts that seem to be the very antithesis of Clausewitz's vision for conducting a successful war.

Restoring Clausewitz: What's at Stake?
Given the strategic failures that have marked American foreign policy, one might reasonably ask whether another book on Clausewitz is necessary. Considering the many in-depth studies on Clausewitz by scholars such as Christopher Bassford, Michael Howard, Peter Paret, Jon Sumida, and

Donald Stoker, published across academic and popular presses, the answer might appear to be "no." It is difficult to argue that policy missteps stem from a lack of access to Clausewitz's writings or to rigorous academic discussion of his ideas. Rather, many of these works have struggled to engage non-specialist readers. They often assume prior expertise, fail to address common misperceptions directly, offer limited historical context, and rely heavily on academic or military jargon. While Clausewitz's writing is inherently challenging, the broader reading public has not been well served by the academic community.

The goal of this book is not to generate new theories, but to synthesize existing scholarship into a more accessible and usable format. Consider, for example, the case of Apple. While the company did not invent the computer or the mobile phone, it achieved tremendous success by making these technologies more intuitive and user-friendly for the general public. In a similar vein, this book aims to "restore" Clausewitz by making his works more accessible and useable.

This book seeks a careful balance—it must be accessible without becoming simplistic. While it is not a beginner's guide, neither is it intended as an academic monograph. Rather, it is meant as a companion to *On War*, offering essential context and interpretive clarity for a broad audience of readers. For those who will never purchase another book about Clausewitz, this book is designed to make his work more user friendly. For readers unlikely to pursue further scholarship on Clausewitz, this volume seeks to make his work more approachable. For those eager to delve deeper, the endnotes offer extensive citations to guide further engagement with the academic literature. In short, this book is intended both as a useful point of entry for new readers and as a constructive guide for those revisiting Clausewitz with fresh questions.

How This Book Engages Clausewitz and *On War*

This book aims to make Clausewitz's most influential work, *On War*, accessible and relevant to an educated but non-specialist audience. To that end, the chapters that follow provide historical context, explain key concepts, and offer interpretive guidance to help readers navigate the text's complexity. Rather than reducing Clausewitz's arguments to simple formulas, the book invites readers to engage critically with his ideas and develop their own insights from this unfinished and complex work.

Chapter 1 explores why Clausewitz's ideas are so frequently misunderstood. It shows that *On War*—unfinished at the time of Clausewitz's death—has been hindered by flawed translations, politicized in interpretation, and remains conceptually elusive. By showing how this work has been misused, this chapter aims to help readers understand and avoid these common pitfalls.

Chapters 2 through 5 offer a biographical account of Clausewitz's military career and intellectual development. Chapter 2 focuses on Clausewitz's early life, beginning with his birth in 1780 and continuing through the outbreak of war with France in 1806. Chapter 3 covers the period from the twin battles of Jena and Auerstädt in 1806 to Clausewitz's marriage in 1810. Chapter 4 examines his decision to leave the Prussian army and serve with the Russian army, concluding with the end of the Napoleonic Wars in 1815. Chapter 5 addresses the final sixteen years of Clausewitz's life, during which he confronted a series of professional setbacks by devoting himself to the manuscript that would later be published posthumously as *On War*. The common theme across these chapters is that Clausewitz's writings were profoundly shaped by his personal experiences. Although often frustrated by his circumstances, he channeled his intellectual energy into his work. Understanding this context allows for a deeper appreciation of his text.

Chapter 6 traces the impact of *On War* and asks why Clausewitz was ignored in his own time yet venerated in ours? The answer to this question is surprisingly straightforward, though rooted in unfortunate historical circumstances. Clausewitz died before completing his manuscript and, unlike contemporaries such as Antoine-Henri Jomini, was not a self-promoter.[9] Moreover, *On War* is not a straightforward checklist but rather a work of philosophy—an approach that made it less appealing to military institutions seeking prescriptive guidance. By tracing the intellectual legacy of Clausewitz's work, this chapter demonstrates how later thinkers and practitioners have drawn on Clausewitz to shape their strategic thinking.

Chapter 7, the first of three theoretical chapters, explores Clausewitz's core concepts concerning the nature of war—political aims, friction, and the trinity—and provides an accessible overview of these foundational ideas. In doing so, the chapter seeks to deepen understanding of these frequently misunderstood concepts.

Chapter 8 continues the theoretical discussion by examining Clausewitz's views on the evolving character of war. It explores how his theories on people's wars, the primacy of the defense, the center of gravity, and the culminating point of the attack reflect his efforts to model the changing conduct of warfare he observed during his lifetime. Like the previous chapter, this section aims to present these key concepts in a manner that is accessible and relevant to contemporary strategic debates.

Chapter 9, the final theoretical chapter, focuses on Clausewitz's conception of military genius. He developed this understanding in response to Napoleon Bonaparte—a figure for whom he held in both admiration and contempt. This chapter examines the twin habits of mind that constitute military genius—*coup d'œil* and determination—and argues that these leadership qualities remain highly relevant today. To apply Clausewitz's model to a modern case, this chapter briefly considers the career of General George Patton, illustrating how his discipline and resolve embodied the qualities of *coup d'œil* and determination as described by Clausewitz.

The conclusion affirms the enduring value of studying Clausewitz. It highlights the complexity of his thought and offers practical guidance for approaching his work with clarity and purpose. It concludes with a deliberately open-ended invitation to engage with Clausewitz as a lifelong intellectual journey—one that promises both challenge and insight. Ultimately, this book seeks to restore Clausewitz to meaningful relevance—a process that is both demanding and deeply rewarding.

Notes

1. Bellinger, *Marie von Clausewitz*, 1.
2. Coker, *Rebooting Clausewitz*, xiii; Bassford, *Clausewitz in English*, 4; Echevarria, *Clausewitz and Contemporary War*, 7; and Fleming, *Moonraker*, 110. For a detailed discussion of Clausewitz's influence on Lenin, see Aron, *Clausewitz: Philosopher of War*, 267–277; Blainey, *The Causes of War*, 153; and Heuser, *Reading Clausewitz*, 19 and 46.
3. Dylan, *Chronicles*, 41 and 45.
4. Clausewitz's work has even been adapted as the basis for works on business strategy. See Holmes, *Carl Von Clausewitz's On War*; Paley, *Clausewitz Talks Business*; and Wallace, *Carl von Clausewitz, The Fog of War, and the AI Revolution*.
5. Paret, *The Cognitive Challenge of War*, 142.
6. For a theoretical understanding on how popular culture shapes perceptions of real-life events, see Daniel and Musgrave, "Synthetic Experiences," 503–516.
7. *Crimson Tide*, dir. Tony Scott.
8. *Crimson Tide*.
9. Jomini read Clausewitz's work with interest and later accused him of appropriating his ideas about war, though there is little reason to believe this claim. If anything, Jomini appears to have adapted his own ideas after reading Clausewitz's work, as suggested by his expanded discussions on the importance of politics for decision-makers. Echevarria, *Clausewitz and Contemporary War*, 15 and 17; and Bassford, *Clausewitz in English*, 50.

CHAPTER 1

CLAUSEWITZ MISUNDERSTOOD: THE CASE FOR RESTORATION

Clausewitz's masterpiece, *On War*, is complex, challenging, protean, and elusive. It has been described as an intellectual Rorschach test—open to interpretation and prone to misreading.[1] It is frequently quoted—often misquoted—and seldom fully understood. It is a work that many begin but few complete, and it often functions more as an impressive-looking talisman than as a fully read text. It is unevenly written and insufficiently edited. It contains internal contradictions. Its length poses a challenge to many readers.[2] In many respects, it reflects the period of Napoleonic warfare in which it was written. Yet, despite limitations, *On War* endures as one of the most influential and enduring philosophies of war ever written.

An Unfinished Masterpiece
One of the most striking revelations for those new to Clausewitz is that his major work, *On War*, remained incomplete at the time of his death. When Clausewitz died suddenly in 1831, he had not worked on the manuscript for nearly a year. He had previously described the unfinished manuscript

as a "shapeless mass of ideas...liable to endless misinterpretation" and "nothing but a collection of materials from which a theory of war was to have been distilled."[3] While his exact plans for revision remain unknown, had he lived to complete the work, it might have been shorter, more cohesive, and more internally consistent, with fuller integration of his insights about people's wars. Although it is tempting to speculate on what might have been, the simple fact remains that *On War* is unfinished.

Fortunately, all was not lost as Clausewitz's wife, Marie, was an exceptional woman in her own right. Marie was highly intelligent and well-educated, and—contrary to the conventions of the time—Carl treated her as an intellectual equal. He valued her wit and intelligence, and their correspondence makes clear that they frequently exchanged ideas about art, politics, literature, and strategy.[4] Much of *On War* was reportedly composed in her bedroom, where the couple felt free to relax and exchange ideas.[5] Even in her grief, Marie saw the publication of this work as a means of preserving and promoting her husband's intellectual legacy. She set about editing the manuscript and finding a publisher almost immediately after his death. Thanks to her extraordinary efforts, *On War* was preserved for posterity, and Carl von Clausewitz was spared historical obscurity.

The text that Marie edited and ultimately published remained unfinished, a fact that has provoked significant debate among Clausewitz scholars. Some scholars point to an undated note left by Clausewitz and conclude that only Book I, Chapter 1 was complete at the time of his death. Others argue that *On War* was essentially complete at the time of his death and, despite the need for editing, sufficiently developed for a careful reader to grasp his core arguments and theoretical insights.[6] This debate remains unresolved. Modern readers are thus left to choose: reject the work as unusable or engage with what remains.

If readers accept that the text is usable, they must avoid the secondary trap of attempting to complete Clausewitz's work for him. It is tempting to assume that further revisions to the text's inconsistencies or elaborated on topics such as guerrilla wars, economics, or naval power. However, a careful reader must accept that we will never truly know what Clausewitz's final version would have contained. Although Clausewitz left a few notes, he never specified his final intentions. He was a perfectionist—perhaps even a tortured genius—and it is quite possible he would never have completed the work. Throughout his life, he left hundreds of memos and papers unfinished, suggesting that he might have never been satisfied with his manuscript no matter how long he lived.[7]

To use *On War*, responsibly, one must accept it as a flawed masterpiece, one that offers only tantalizing glimpses of what Carl von Clausewitz may have ultimately believed.

A Philosophy of War

What, then, did Clausewitz hope to accomplish with his book *On War*? In short, he sought to write a philosophy of war that would endure through the ages.[8] Clausewitz was profoundly influenced by the continental philosophers of his day, most notably, Immanuel Kant, Fredrick Hegel, Johann Kiesewetter, Johann Gottlieb Fichte, Johann Gottfried von Herder, and Johann Wolfgang von Goethe.[9] Although he found much to disagree with in their works (Kant's deontological ethics and his *Perpetual Peace* to name just two), he sought to be taken seriously as an intellectual and therefore adopted many of their academic and literary conventions.[10]

Of these academic conventions, none was more central to Clausewitz's writings than the dialectic method adapted from Kant and Hegel.[11] This method of exploring complex ideas was popular because, as Hegel famously argued, "a sentence in the form of a judgment is not the best way of expressing speculative truths."[12] Rather than offering a simple declaration of truth, this method tests a purportedly true statement by subjecting it to its logical alternatives.[13]

This method resembled a debate between two opposing views and developed ideas through three distinct phases: thesis, antithesis, and synthesis.[14] In the thesis section, one position is advocated, which is then contradicted by the antithesis.[15] The thesis and the antithesis thus set the extreme boundaries of the debate.[16] The two opposing positions are then reconciled in a synthesis. The synthesis is critical because it moves the argument closer to an objective truth by resolving the logical flaws of both the thesis and the antithesis.[17] Once the synthesis has been formed, it becomes the basis for a new thesis, which is then countered with a new antithesis, leading to a new synthesis. Ideally, this process is repeated *ad infinitum* to achieve greater rigor, logic, and proximity to objective truth. Clausewitz defended this method of separating ideas and then reassembling as "necessary...if theory is to serve its principal purpose of *discriminating between dissimilar elements* (emphasis in the original)."[18]

Figure 1. The Dialectic.

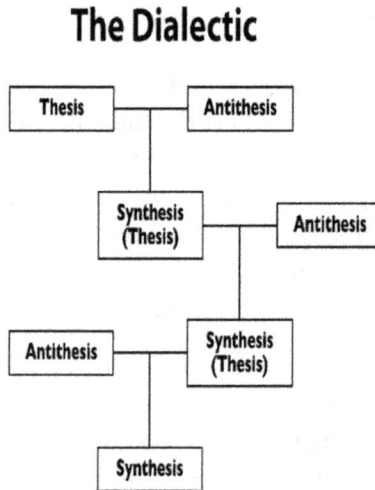

The Dialectic

```
┌──────────┐       ┌────────────┐
│  Thesis  ├───┬───┤ Antithesis │
└──────────┘   │   └────────────┘
       ┌───────┴────┐
       │ Synthesis  │       ┌────────────┐
       │ (Thesis)   ├───┬───┤ Antithesis │
       └────────────┘   │   └────────────┘
                        │
┌────────────┐   ┌──────┴─────┐
│ Antithesis ├───┤ Synthesis  │
└────────────┘   │ (Thesis)   │
            ┌────┴───────┐
            │ Synthesis  │
            └────────────┘
```

While the dialectical method can fruitfully reconcile opposing views, expose flaws in reasoning, and reveal essential truths, it does not lend itself to quick comprehension. Rather, many find it pedantic, overly formalized, confusing, and even tedious. As one scholar noted, the rapid point-counterpoint style can be exhausting, evoking "sea-sickness" as Clausewitz rapidly alternates between opposing ideal types.[19] This is particularly true in Book I, Chapter 1—the only section Clausewitz considered finished—which is often the first (and often only) part of *On War* that readers encounter before becoming frustrated, intimidated, and confused.

More problematic still is that since this method presents extreme positions before reconciling them, readers must recognize that Clausewitz considers *both* positions flawed. A careless reader may easily overlook this and mistake the thesis or the antithesis for Clausewitz's actual argument. Worse still, these deliberately extreme formulations make it easy to cherry-pick quotations and falsely attribute extreme views to Clausewitz.[20] One famous example from *On War* (discussed in greater detail later) is Clausewitz's thesis that violence in war expands until it reaches its utmost extremes. He then counters with the antithesis that war is never truly extreme. Clausewitz then offers the synthesis: war can tend toward extremes, but policymakers must ensure it never becomes truly extreme.

This is a brilliant formulation—but also deeply confusing. Indeed, many readers have either inadvertently missed these distinctions or deliberately focused on Clausewitz's extreme theses or antitheses, neglecting the more moderate and truthful syntheses. As Clausewitz explained:

> Once again we must remind the reader that, in order to lend clarity, distinction, and emphasis to our ideas, only perfect contrasts, the extremes of the spectrum, have been included in our observations. As an actual occurrence, war generally falls somewhere in between, and is influenced by these extremes only to the extent to which it approaches them.[21]

As if the complexity of the philosophical style were not enough, Clausewitz compounded the difficulty within his own writings.

Unlike many philosophers who employed this method, Clausewitz did not always insist on resolving his dialectic discourses. Instead, he used the method to introduce and define opposing concepts, but often left them unresolved.[22] Whether this was intentional or simply a consequence of the work's incomplete state is unclear, but the text as it stands departs from a purely dialectic treatment. Clausewitz frequently departs from strict dialectical structure by employing the rule of three throughout his work. While this convention works well for describing concepts, such as his famous trinity, it differs fundamentally from the dialectic, as it illustrates the relationship among three elements rather than two. The incomplete status of the book further complicates this issue as Clausewitz's use of the rule of three is often incomplete or confusing. For example, in Book V, Chapter 17, he discusses the three types of terrain but lists four, "mountains, forests and marshes, and agricultural areas."[23] Why? Was he suggesting that forests and marshes are the same or pose similar military challenges? Was this an awkward attempt to force the rule of three? Or simply a logical or editorial oversight? Ultimately, we do not know.

In addition to this dialectical method, Clausewitz also developed his own form of critical analysis (*Kritik* in the original German).[24] This method of counterfactual reasoning sought to apply logic to the study of history as a means of testing theories and revealing deeper truths. Clausewitz believed that by studying the information, intentions, and means available to past commanders, one could develop a sympathetic understanding of why they made certain decisions, what the consequences were, and how they might have acted differently. Here again, Clausewitz sought to follow in the intellectual footsteps of the Enlightenment philosophers who had inspired him and to apply their methods to the study of war.

To construct a rigorous and comprehensive set of counterfactuals, Clausewitz introduced considerable complexity into his own work. Given that war is a competitive and dynamic enterprise, Clausewitz needed to develop more than a fixed understanding of a single commander at a single moment. Rather, this method required a sympathetic understanding of the decision makers at all levels of command, on all sides of a conflict, and at multiple stages of a campaign as the fortunes of war shifted and new information emerged. Even for a thinker as dedicated to his craft as Clausewitz, this was an almost impossible task. By his own admission, his method "may lead to a broad and complex field of inquiry in which we may easily get lost," but it was the only way to begin to understand the complex and contingent nature of war.[25]

For the modern reader, this method is at once familiar and challenging. On the one hand, Clausewitz's approach resembles the case-study method employed by modern social scientists and the staff-ride method used by militaries worldwide.[26] Put simply, the goal is to use the past to explain significant events, test assumptions, and anticipate how future outcomes might be shaped. On the other hand, this approach can appear academic, remote, and impenetrable. For modern readers, it is often difficult to grasp the historical context of nineteenth-century warfare well enough to interpret Clausewitz's references to specific leaders, battles, treaties, and social movements. Further complicating this challenge is Clausewitz's tendency to adopt the perspective of commanders such as Napoleon and to shift fluidly between their viewpoint and his own. As a result, it is often difficult to distinguish between Clausewitz's reconstruction of what Napoleon may have thought at a particular moment and Clausewitz's own interpretation with the benefit of hindsight. This methodological approach should prompt caution as it is critical to understand *how and why* Clausewitz employs a particular historical example. While these counterfactual examples may be rewarding for readers with the knowledge and time to engage with them fully, they often appear, on the surface, as meandering and overly technical campaign studies with few clear or compelling lessons.

The challenge for the reader is to study Clausewitz's writings carefully and to accept the fact that they are at times contradictory and confusing without imposing personal interpretations or biases onto his work.[27]

Rather than offering simple answers, Clausewitz's philosophy demands careful reading and rereading.

The Challenges of Translation and Meaning

Another challenge for the modern reader is that Clausewitz often selected words with complex, layered meanings. For example, consider the word *Politik*—a word that lacks a precise English equivalent.[28] Clausewitz's *On War* centers on politics, and its most quoted line is that "war is an extension of politics by other means."[29] Therefore, it may seem straightforward to translate *Politik* as "politics." However, this would be a mistake. The word "politics" is itself a complex concept with multiple potential meanings. Does it refer to party politics, ideology, international relations, local politics, bureaucratic politics, policy, or the business of a *polis*? "Politics" is also a loaded term that elicits powerful emotional responses from readers. Indeed, like war, politics is frequently a messy and imprecise endeavor.

While Clausewitz considered each of these elements to some degree, it is generally agreed that he used the word *Politik* to mean "policy." When understood in this way, this word transforms one's interpretation of *On War*. Clausewitz was not writing about ideology or partisan politics but rather about something more limited and rational. A war that serves a policy end is not waged on a whim or in pursuit of transformative philosophical principles. Rather, a war conducted for policy is designed to achieve a specific objective and should be restrained in its purpose.[30] Although the imperative to wage war rationally is central to Clausewitz's work, a casual reader might overlook this point—particularly because of the ambiguity inherent in the word "politics."[31]

For a similar case of a term with a complex meaning, consider Clausewitz's use of the word *Volk*. This term literally means "people," but Clausewitz also uses it to signify "nation."[32] This broader interpretation is significant, as Clausewitz was a pan-German nationalist writing during a period when peoples and their governments were beginning to merge into modern nation-states. Given that Clausewitz roughly equates the people, the military, and the government as the three components of his trinity, it is essential to grasp both their distinctions and their interrelationships. Once, again, Clausewitz's own word choices complicate an already difficult analytical task, making it essential to approach them carefully and with a sound understanding of their historical context.[33]

If these subtle word choices were not confusing enough, Clausewitz also frequently used multiple terms interchangeably to express the same concepts. Had he lived to complete his work, he might have resolved these discrepancies, which introduce yet another layer of inconsistency and difficulty. For example, Clausewitz's work appears inconsistent with its use of the words *Ziel* and *Zweck*. *Ziel* means "goal" or "objective," whereas *Zweck* means "purpose" or "point." At various points in Books I, II, VII, and VIII, Clausewitz uses these terms in ways that appear to contradict one another. This inconsistency poses a challenge for translators, especially in a text whose nuances compel readers to scrutinize every word of key passages. Many later translators have had to choose between preserving Clausewitz's inconsistencies or exercising their judgment to impose greater consistency.[34]

The difficulty of translating this dense text, combined with Clausewitz's relative obscurity in the English-speaking world, meant that no suitable English translation appeared until the last quarter of the twentieth century. Despite renewed interest in the Prussian military following the Franco-Prussian War of 1870–1871, and quotations about Clausewitz by General Moltke himself, the first English translation by J. J. Graham sold only thirty-nine copies by 1880.[35] As a result of the poor translation and dismal sales, Clausewitz remained virtually unknown in Britain and the United States. Although his works attracted considerable interest among a small circle of British military intellectuals, his influence on the broader English-speaking world remained minimal, due in part to these persistent translation issues.

Politicization, Personalization, and Purposeful Mischaracterization

Unfortunately, the lack of accurate translations of *On War* is not the only factor contributing to the widespread misunderstanding of Clausewitz. Generations of critics have compounded the problem by perpetuating the myth that he was merely a warmongering Prussian whose ideas drove successive generations of German leaders to violence and destruction. While a hurried reading can easily lead to misunderstandings of complex arguments, it is irresponsible to claim that Clausewitz advocated for unrestrained violence or directly inspired leaders such as Kaiser Wilhelm II or Adolf Hitler.

Perhaps no single writer has done more to distort the popular under-standing of Clausewitz's ideas than the British journalist and historian Basil Liddell Hart.[36] Liddell Hart was a prolific and influential writer whose career spanned six decades. He remains a controversial figure for advocating appeasement during the interwar period, for promoting the myth of the apolitical and "clean" Wehrmacht, for overstating his role as a pioneer of armored warfare, and for his relentless criticism of Clausewitz. Indeed, despite his mild manners, Liddell Hart appeared to actively seek out controversial positions in his writings, including his interpretation of Clausewitz's ideas.

Though Clausewitz had been dead for more than eighty years when World War I began, Liddell Hart blamed his theories for the carnage in the trenches. He dubbed Clausewitz the "Mahdi of Mass and mutual massacre," an "evil genius of military thought," and the "apostle of total war." According to this argument, the cult of German militarism—and by extension the massed frontal assaults of the First World War—could be traced back to the supposedly poisonous theories of Clausewitz.[37]

If this were not damming enough, Liddell Hart also claimed that Clausewitz's writings had led several key Allied commanders down the path of wasteful attrition strategies. In his 1931 biography of the French general Ferdinand Foch, Liddell Hart argued that Foch had adopted his signature offensive tactics under the influence of *On War*.[38] He portrayed Foch as intellectually simplistic, noting that:

> The ponderous tomes of Clausewitz are so solid as to cause mental indigestion to any student who swallows them without a long course of preparation. Only a mind developed by years of study and reflection can dissolve the solid lump into digestible particles. Critical power and a wide knowledge of history are also necessary for producing the juices to counteract the Clausewitzian fermentation.[39]

According to Liddell Hart, Foch used Clausewitz to reinforce his personal will and his inclination toward aggressive action. Because of this hasty misreading, "Foch's main contribution to the French theory of war was to strengthen its Clausewitzian character," and the slaughters of the Great War were, in Liddell Hart's view, the tragic but logical result.[40]

According to Liddell Hart, British failures in the First World War also stemmed from "the military mode of thought inspired by Clausewitz." He wrote:

> That we adopted it is only too clear from analysis of our pre-war textbooks, of the strategical memoranda, drawn up by the General Staff at home and in France during the war, and of diaries and memoirs of the dominant military authorities published since the war. They are full of tags that can be traced to Clausewitz, if often in exaggerated transfer.[41]

Liddell Hart's claim is highly irresponsible. He never specifies whose "diaries or memoirs" he is referencing, nor does he identify the "dominant military authorities" who had "published since the war." He seems to see the ghost of Clausewitz everywhere, tracing vague "tags" to him through exaggerated associations. In short, these sweeping accusations lack even the slimmest shred of supporting evidence.[42]

There are many issues with this interpretation. First, Clausewitz's works were not widely read before or during the Great War, and even within the German military, he was regarded as a relatively minor theorist. Second, while Foch and other Allied generals may have been aware of Clausewitz, there is no evidence that they drew inspiration from his work when planning offensive operations during the war.[43] Rather, the adoption of offensive strategies appears to have been a practical effort to foster fighting spirit and unify a deeply divided French military in the wake of the Dreyfus Affair, not the result of a distorted Clausewitzian doctrine.[44] Third, this interpretation ignores the fact that Clausewitz advocated battle and mass attacks only when they offered a reasonable prospect of decisive results; he would not have endorsed wasteful frontal assaults.[45] Finally, the most influential theorist before the outbreak of World War I was Clausewitz's rival, Henri Jomini, who enjoyed a far wider readership and placed greater emphasis on mass attacks against critical points.[46]

Unfortunately, Liddell Hart appears to have misunderstood both the historical context and Clausewitz's work. Shaped by his traumatic experiences during the Great War, he developed a strong anti-German bias and was quick to assign blame to Clausewitz in his early writings. As his reputation grew, Liddell Hart attributed to Clausewitz an ever-expanding list of ills, including German revanchism and strategic failures during World War II. Despite the emergence of more rigorous scholarship, he seems never to have reconsidered his early claims.[47] It would be easy to dismiss Liddell Hart as a polemicist or a careless scholar were it not for his stature as a prolific and influential writer of military history. Personally known for his gracious manner, Liddell Hart mentored generations of British and American academics. His combination of commercial success and scholarly reach gave him outsized influence over both the field of military history and the popular understanding of Clausewitz.[48]

A similar example of anti-Clausewitz bias in British scholarship appears in J. H. Morgan's 1945 book, *Assize of Arms*. Morgan, a prominent international lawyer and adviser to the Nuremberg war crimes tribunals, gained recognition for exposing the atrocities of the Nazi regime. He argued that the roots of Nazi brutality lay in Clausewitz's ideas, which he claimed advocated the mistreatment of prisoners, the pursuit of maximum destruction, the glorification of war as "the highest goal of human achievement," and the "necessity of terrorism." According to Morgan, Clausewitz was a brute because "the sight of the bloodiest battlefields left him not only quite unmoved but exultant." This cruelty , he asserted, had permeated the entire German military as, "Clausewitz, a typical Prussian, indoctrinated the same German military thinkers with the 'philosophy' of brutality."[49]

Although Morgan's claims were hyperbolic and irresponsible, they resonated with a British public that had suffered immensely during the two world wars. These conflicts not only inflicted millions of casualties, dismantled the British Empire, and precipitated fiscal collapse, but also left a lasting imprint on literature and public memory.[50] During both world wars, British propaganda had vilified Clausewitz's name as a proxy for German militarism and brutality, creating a receptive and war-weary audience for Morgan's arguments.[51]

As a moral crusader rather than a trained scholar, Morgan's mistaken assumption that Clausewitz was an influential figure in Nazi Germany is perhaps understandable. Although the Nazi regime misappropriated Clausewitz's legacy for its own purposes—more than a century after his death—his ideas were not widely disseminated within the German military at the time, and it is unclear whether Hitler or other senior figures had anything more than a superficial understanding of his theories. Morgan was also mistaken in asserting that *On War* advocated for maximum suffering and destruction; this conclusion appears to have stemmed from his personal experience prosecuting Nazi atrocities rather than from a careful scholarly reading.[52] Unfortunately, this caricatured interpretation persisted in the postwar period, leading to the continued mischaracterization of Clausewitz's ideas.

A final prominent but misguided critic of Clausewitz was the British historian John Keegan.[53] Keegan achieved considerable fame for his

1976 book *The Face of Battle*, which advocated a bottom-up approach
to military history. This groundbreaking work established Keegan as
one of the most influential military historians of his era and gave him
a platform to comment on a wide range of issues concerning the use
of force. Keenly aware of the profound human cost of modern warfare,
Keegan adopted a series of idealistic positions that called for moving
beyond warfare as an instrument of statecraft.[54]

To this end, Keegan took direct aim at Clausewitz's view that war is
a political act. In his influential 1993 book *A History of Warfare*, Keegan
argued that warfare is social, not political, and that a Clausewitzian
understanding of conflict is outdated, dangerous, and immoral.[55] Like
Liddell Hart and Morgan, Keegan repeated the trope that Clausewitz
was an influential warmonger whose ideas fueled two world wars and
led to untold suffering:

> Clausewitz was the apostle of a revolutionary philosophy of
> warmaking, which sought to depict war as a political activity to a
> caste that held politicians to be anathema...Clausewitz conceived
> a theory which elevated the regimental officer's values—total
> dedication to duty, even to dying in the cannon's mouth—to the
> status of a political creed, thereby absolving him from deeper
> political reflection.[56]

Keegan's criticisms were ill-informed and exaggerated, but they were
nevertheless influential because of his status as a popular historian.
Although Clausewitz lamented the cruelty of war and was deeply reflec-
tive, this reality is absent from Keegan's portrayal.[57]

As discussed in detail in the biographical sections of this work, Clausewitz understood the horrors of war because he had experienced them
firsthand. While *On War* affirms that war can serve as an instrument
of statecraft, it also argues for restraint and care in its application.[58] To
ignore these facts is to perpetuate a caricature of Clausewitz as little
more than a Prussian bogeyman.

The Modernist Case against Clausewitz

Another line of criticism argues that Clausewitz's works are outdated and obsolete.[59] According to this view—advanced most prominently by Anatol Rapoport, Martin van Creveld, and Mary Kaldor—Clausewitz's theories are an irrelevant anachronism because the nature of war and the state system have fundamentally changed since his time.[60] In particular, Clausewitz's state-centric theory cannot account for the transformation of warfare in the decades following World War II, during which non-state actors have risen in importance and the state's monopoly on violence has eroded. This argument holds that Clausewitz's theories will continue to fade into irrelevance as the world returns to a pre-Westphalian security order. In such a world, a mix of local power brokers, criminal networks, religious authorities, and corporations would replace nations as the primary security providers. With the collapse of the state system, the unwritten trinitarian social contract described in *On War* will fade from existence, along with the last vestiges of Clausewitz's relevance.[61]

At first glance, this line of reasoning may appear superficially plausible. Non-state violence has become more prevalent in the post–World War II era, and nation-states have had a mixed record of success in combating these seemingly weak opponents. The power of substate groups has been amplified by technologies such as computers, social media, drones, and mobile phones. The computing power and firepower now available to billions of ordinary citizens have reached levels almost unimaginable a century ago, enabling nonstate actors to compete on a more equal footing. These emerging technologies have clearly challenged the hold of nation-states on power and compelled them to respond to these new threats.

Despite these developments, the rumors of the demise of nation states and trinitarian warfare appear greatly exaggerated. Historically, nation states have proven remarkably durable, and there is little reason to doubt that they will remain the dominant actors in international affairs for decades to come.[62]

When considering these challenges, it is important to recognize that even the smallest nation dwarfs the size and power of super-empowered individuals or violent groups. Moreover, most actors that challenge nation-states do not seek to topple the entire state system, but rather to join and reshape it from within. Even groups like ISIS or anti-Chinese protestors seek legitimacy within the existing system by seizing power and establishing a state government aligned with their political agendas. Rather than signaling the demise of the nation-state and trinitarian warfare, this suggests that the threatened use of violence remains integral to the political process—just as Clausewitz predicted.

Another common shortcoming of these modernist critics is their failure to fully understand Clausewitz's work. While the most frequently cited sections of *On War* focus on state-to-state warfare, substantial portions address revolutionary violence and people's wars. These sections are often overlooked because they are less quotable and appear in the less frequently read, more roughly drafted middle chapters of the book; nevertheless, Clausewitz clearly understood the importance of non-state actors and sought to incorporate them into his theory. Indeed, Clausewitz had been thinking seriously about people's wars as early as 1810, when he taught a course on the subject at the Prussian War College. There is strong evidence that he intended to incorporate his views on the subject more fully into the manuscript had he lived to complete it. Given this, it appears that those who claim Clausewitz failed to grasp the significance of non-state warfare possess, at best, a shallow understanding of his life and work.[63]

Finally, these modernist critics fail to recognize that Clausewitz's work has endured precisely because it does not attempt to serve as a tactical or operational guide to war. Rather, it offers a theoretical framework for understanding war. By highlighting the changing elements of war's character while asserting that its nature remains constant across time, Clausewitz's work retains inherent flexibility. While tactics and technology will inevitably change, Clausewitz maintained that, as a human endeavor, war's essential nature remains unchanged.

Overall, many of Clausewitz's most prominent critics have approached his work in ways that are unfair, biased, and ahistorical.[64] Although these criticisms, in some ways, kept his name in circulation, they have also distorted his legacy and perpetuated falsehoods and half-truths. Many of these misconceptions have proven remarkably persistent and continue to be repeated today. The net result of these unfair criticisms is that they hinder a proper understanding of Clausewitz's work.

Embracing Complexity and Understanding Context

So how does one begin to understand a work that is incomplete, seemingly contradictory, difficult to translate, and has been politicized and caricatured by generations of scholars? The answer is uncomfortable—embrace the complexity and understand the context.[65]

Embracing complexity is inherently difficult. Human beings are, in fact, accustomed to thinking in relatively simple terms. If one drops an object, it falls. If one waters a plant, it grows. If one eats better and exercises, one will (albeit more slowly with age) lose weight. Although each of these phenomena is more complex than it appears, our basic understanding of gravity, plant growth, and metabolism generally follows a simple cause-and-effect logic. This level of understanding suffices such that these ideas rarely require sustained reflection or debate. They hold true for most people, most of the time.

Clausewitz's conception of war is markedly different. He believed war to be the most complex of all human undertakings and, as such, one that defies simple theorizing. He wrote that war was so complex that:

> Newton himself would quail before the algebraic problems it could pose. The size and variety of factors to be weighed, and the uncertainty about the proper scale to use are bound to make it far more difficult to reach the right conclusion.[66]

This was more than mere epistemological humility; Clausewitz had studied war for nearly his entire life and participated in dozens of battles. Yet even with this extensive experience, he believed he could only begin to describe war's fundamental elements. To that end, he explicitly rejected reductionist approaches and embraced concepts such as nonlinearity and chaos.[67] In short, he recognized that war defied simple explanation and constructed a theory to account for its extraordinary complexity. Although challenging, it remains essential to engage with Clausewitz on his own terms and embrace the complexity of his approach.

As we seek to engage with these complex theories, it is also critical to develop a sympathetic understanding of Clausewitz's historical context. As the next four chapters will demonstrate, Clausewitz's work was profoundly shaped by the rapidly changing world in which he lived. He witnessed the French Revolution and the Napoleonic Wars firsthand—and was both horrified and inspired. To that end, he developed complex theories that resisted quick or convenient answers, seeking instead to make sense of his lived experiences. While it may seem obvious that Clausewitz's personal context matters, this fact has been too often been overlooked in analyses of his work. As Peter Paret, one of Clausewitz's most careful biographers, observed, "Every theory that outlasts its creator tends to be reinterpreted unhistorically...[but] Clausewitz's writings have suffered the attendant distortions more than most."[68]

The remainder of this book will engage with the complexity of Clausewitz's theories while tracing the interplay between his intellectual development and the upheavals of his time.[69] It is to this vital context—the source of Clausewitz's theories—that we now turn.

NOTES

1. For an argument that "*On War* obligingly reflects back the intellectual predispositions of whoever look into it," see Alterman, "The Uses and Abuses of Clausewitz," 18–32; and Bassford, *Clausewitz in English*, 9.
2. Coker, *Rebooting Clausewitz*, xi.
3. Quoted from "Note of 10 July 1827" and "Unfinished Note, Presumably Written in 1830" respectively. Clausewitz, *On War* (Howard and Paret), 70.
4. Bellinger, *Marie von Clausewitz*, 4, 6, 10, and 36.
5. Strachan, *Clausewitz's On War*, 66.
6. Sumida, *Decoding Clausewitz*. For a critique of Sumida's interpretation, see Paret, *The Cognitive Challenge of War*, 118–119.
7. The incompleteness of *On War* has been interpreted by some as evidence of Clausewitz's rigorous academic mind; see Strachan, *Clausewitz's On War*, 104 and 193.
8. "On the Genesis of his Early Manuscript on the Theory of War, Written around 1818" in Clausewitz, *On War* (Howard and Paret), 63.
9. Echevarria, *Clausewitz and Contemporary War*, 3–4, 9, 22–24, 37–38, 47, 110–112, 127, and 155. See also Beiser, *Enlightenment, Revolution, & Romanticism*, 189–260; Craig, *The Germans*, 190–212; Paret, *The Cognitive Challenge of War*, 140; Paret, *Clausewitz and the State*, 69 and 75, and Strachan, *Clausewitz's On War*, 90–91.
10. Lebow, *The Tragic Vision of Politics*, 214–215; and Strachan, *Clausewitz's On War*, 90. For a more technical analysis comparing Clausewitz to Kant and Hegel, see Aron, *Clausewitz: Philosopher of War*, 223–230. For a counterpoint that emphasizes their shared methodology but differing conclusions, see Cormier, "Hegel and Clausewitz," 419–444.
11. The first known appearance of the dialectic in Clausewitz's writing dates to 1815; see Lebow, *The Tragic Vision of Politics*, 188.
12. Andreas Herberg-Rothe, *Clausewitz's Puzzle*, 118
13. Strachan, *Clausewitz's On War*, 90.
14. On purely technical grounds, Clausewitz's work does not employ a dialectic method as he rarely used true opposites. For example, a strict application of dialectics would contrast attack with non-attack or defense with non-defense, whereas Clausewitz contrasts offense with defense; see Herberg-Rothe, *Clausewitz's Puzzle*, 135–138.

15. For a discussion of Clausewitz's use of pairs of contrasting ideas, see Herberg-Rothe, *Clausewitz's Puzzle*, 3–4 and 119–138; and Waldman, *War, Clausewitz, and the Trinity*, 31–44.

16. Herberg-Rothe, *Clausewitz's Puzzle*, 117.

17. Lebow, *The Tragic Vision of Politics*, 188.

18. Clausewitz, *On War* (Howard and Paret), II.1, 129.

19. Waldman, *War, Clausewitz, and the Trinity*, 36.

20. Echevarria, *Clausewitz and Contemporary War*, 37.

21. Clausewitz, *On War* (Howard and Paret), VI.30, 517.

22. Paret, *Clausewitz and the State*, 84

23. Clausewitz, *On War* (Howard and Paret), V.17, 348. See also Strachan, *Clausewitz's On War*, 85–86.

24. For Clausewitz's somewhat convoluted and confusing description of this method, see Clausewitz, *On War* (Howard and Paret), II.5, 156–169.

25. Clausewitz, 159.

26. For modern case-study methodology, see George and Bennett, *Case Studies and Theory Development in Social Sciences*. For the methodology of staff rides, see Robertson, *The Staff Ride*.

27. Strachan, *Clausewitz's On War*, 194.

28. Fleming, *Clausewitz's Timeless Trinity*, 68.

29. Others have argued that the words "by other means" are also misunderstood; see Herberg-Rothe, *Clausewitz's Puzzle*, 5.

30. Strachan, *Clausewitz's On War*, 102–105, 123–124, and 164–165; and Waldman, *War, Clausewitz, and the Trinity*, 81–101.

31. Waldman, *War, Clausewitz, and the Trinity*, 85.

32. Strachan, *Clausewitz's On War*, 165.

33. For a further discussion about the meaning of other terms such as "maneuver," see Aron, *Clausewitz: Philosopher of War*, 74–81.

34. Aron, *Clausewitz: Philosopher of War*, 61 and 113; Howard, *Clausewitz*, 35–36; and Strachan, *Clausewitz's On War*, 102-105.

35. Bassford, *Clausewitz in English*, 57–58 and 70; and Willmott and Barrett, *Clausewitz Reconsidered*, 23.

36. Aron, *Clausewitz: Philosopher of War*, 125, 127, 222, 233–234, 237, and 278–280.

37. Bassford, *Clausewitz in English*, 129–130 and Sumida, *Decoding Clausewitz*, 25–35.

38. Strachan, *Clausewitz's On War,* 15–16.

39. Liddell Hart, *Foch: The Man of Orléans*, 23; and Luvaas, "Clausewitz, Fuller, and Liddell Hart," 197–212.

40. Liddell Hart, *Foch*, 24; and Porch, "Clausewitz and the French, 1871–1914," 287–302.

41. Liddell Hart, *The British Way of Warfare*, 17; and Bassford, *Clausewitz in English*, 67 and 105–106.

42. Porch, "Clausewitz and the French, 1871–1914," 287–302; and Sumida, *Decoding Clausewitz*, 27–28.

43. Raymond Aron noted that while Clausewitz may have inspired French generals, they learned the opposite lessons from those he intended. As Aron remarks, "Is an author to blame for the folly of those who cannot reason, count or argue?" See Aron, *Clausewitz: Philosopher of War*, 256 and 264.

44. Porch, "Clausewitz and the French, 1871–1914," 287–302. See also Snyder, *Ideology of the Offensive*.

45. Sumida, *Decoding Clausewitz*, 32–33.

46. Liddell Hart, *Strategy*, , 2nd rev. ed., 342. See also Echevarria, *Clausewitz and Contemporary War*, 15. Bassford, *Clausewitz in English*, 106 and 129–130; and Mearsheimer, *Liddell-Hart and the Weight of History*.

47. Luvaas, "Clausewitz, Fuller, and Liddell Hart," 197–212; and Sumida, *Decoding Clausewitz*, 34.

48. Bassford, *Clausewitz in English*, 130 and 144–151. Clausewitz plays only a small role in the recollections of captured German generals. The only general to mention him was Paul Ludwig Ewald von Kleist, who remarked, "Clausewitz's teachings had fallen into neglect in this generation—even at the time I was at the War Academy and the General Staff. His phrases were quoted, but his books were not closely studied. He was regarded as a military philosopher, rather than a practical teacher. The writings of Schlieffen received much greater attention." See Liddell Hart, *The German Generals Talk*, 194.

49. Morgan, *Assize of Arms*, 71, 119, 179, 249, 250; and Bassford, *Clausewitz in English*, 123–125.

50. Clark, *The Donkeys*; and Fussell, *The Great War in Modern Memory*.

51. Bassford, *Clausewitz in English*, 123.

52. Baldwin, "Clausewitz in Nazi Germany," 5–26. Adolf Hitler appears to have had a selective and distorted understanding of Clausewitz, often quoting *On War* to generals who had little knowledge of it; see Fritz, *The First Soldier*, xiv, xxxii, 5–17, 24, 26, 36, 43, 45–46, 62, 72, 86, 98, 104, 109, 113, 161, 169, 198, 203, 239, 335, 351, 360, 367, 369, 374. For a more dismissive view of Clausewitz's influence on Hitler's worldview and strategy,

see Aron, *Clausewitz: Philosopher of War*, 277–289; and Wallach, "Misperceptions of Clausewitz's *On War* by the German Military," 213–239.

53. Bassford, "John Keegan and the Grand Tradition of Trashing Clausewitz," 319–336.

54. Herberg-Rothe, *Clausewitz's Puzzle*, 157 and 162–164.

55. Keegan makes seventy-four separate references to Clausewitz in his work; see Keegan, *A History of Warfare*.

56. Keegan, 18.

57. For a pointed critique of Keegan's analysis of Clausewitz, see Pinker, *Sense of Style*, 170–171, 178, and 180–183.

58. Lebow, *The Tragic Vision of Politics*, 168 and 215.

59. Van Creveld, *The Transformation of War*; Simpson, *War from the Ground Up*; Smith, *The Utility of Force*; Dimitriu, "Clausewitz and the Politics of War," 645–685; Metz, "A Wake for Clausewitz," 199–213; and Willmott and Barrett, *Clausewitz Reconsidered*. Van Creveld was not always so dismissive of Clausewitz; van Creveld, "The Eternal Clausewitz," in *Clausewitz and Modern Strategy*, 35–50.

60. Huber, "Clausewitz is Dead," 119–121.

61. Kaldor, *New & Old War*.

62. On the stability of the international system, see Waltz, *Theory of International Politics*, 91–92, 134, 162, and 199–204.

63. Strachan, *Clausewitz's On War*, 5–6 and 190.

64. Howard, *Clausewitz*, 65–66.

65. Lebow, *The Tragic Vision of Politics*, 169.

66. Clausewitz, *On War* (Howard and Paret), VIII.3, 586. For a very similar quote, see also Clausewitz, *On War* (Howard and Paret), I.3, 112.

67. Beyerchen, "Clausewitz, Nonlinearity and the Unpredictability of War," 59–90.

68. Paret, *Clausewitz and the State*, 8.

69. Aron, *Clausewitz: Philosopher of War*, 3.

CHAPTER 2

SOLDIER AND SCHOLAR: CLAUSEWITZ'S EARLY DEVELOPMENT

To appreciate Carl von Clausewitz, it is crucial to delve into his biography and understand the formative events of his life.[1] Indeed, nearly every major theoretical insight he developed in *On War* can be directly tied to his lived experiences.[2] In fact, he emphasized the importance of lived experience in *On War*, stating that a " book cannot really teach us how to do anything" and that we must learn by doing.[3] The following four chapters will examine Clausewitz's biography with the goal of providing context and tracing how his lived experiences shaped his theoretical work.

Birth and Early Life
Carl von Clausewitz was born on July 1, 1780, in the Prussian duchy of Magdeburg. His father was a minor Prussian customs official, and he was the fourth and youngest son of six children.[4] Despite their use of the title "von" in their family name, the Clausewitz family's claim to nobility was tenuous.[5] It is unclear if they had ever been granted a title

by royal decree and Clausewitz himself would spend significant energy as an adult ensuring that his noble status was officially recognized by the monarchy. Moreover, the family's chronic financial hardship prevented Clausewitz from enjoying the privileges his title might have implied. This combination of questionable social status and financial insecurity appears to have profoundly shaped the young Clausewitz's life. He lacked the security of societal elites and was deeply self-conscious about his social standing. This tension produced a peculiar mix of ambition to earn distinction and fear that his humble origins, limited education and refinement, and questionable noble status might be exposed.

As a boy, Carl was inspired by his grandfather, a professor of theology, and longed to attend university and pursue an academic path of his own.[6] Unfortunately, his family's limited means rendered this dream of higher education unattainable.[7] Compounding matters, Clausewitz's primary schooling was inadequate, as his family could neither afford elite schools nor private tutors.[8] Lacking a solid formal education, Clausewitz carried a persistent sense of intellectual inferiority throughout his life. For instance, he genuinely feared failing out of the war college and, despite his excellent grades, was astonished to discover that he would not only graduate but was first in his class.[9]

This sense of intellectual inadequacy came to define Clausewitz. He spent the remainder of his life striving to overcome his early educational deficiencies, becoming a lifelong and self-directed learner. Despite these efforts, he remained anxious about appearing foolish in more learned and cultured circles. This deep-seated self-doubt manifested itself in reclusiveness and perfectionism. Clausewitz shared his ideas only with his trusted inner circle and rarely felt confident enough to publish his work under his own name. These traits limited the reach of Clausewitz's ideas during his lifetime and nearly consigned him to total obscurity.

Military Service

As the fourth and youngest son in an era of primogeniture, Clausewitz faced even dimmer financial prospects than his brothers. He understood that his scholarly ambitions were unrealistic given his financial circumstances and ultimately resigned himself to the reality that military service, not education, was his best path to social advancement and self-improvement.[10]

From an early age, Clausewitz was encouraged to seek a military career by his family and friends. His father, Friedrich Gabriel, had briefly served in the army of Frederick the Great but left the service at age twenty-seven after his hand was partially disabled in battle. While his father wanted to continue to serve, he was forced to take a poorly paid position in the Prussian internal revenue service instead. Fredrich romanticized his days as a soldier, maintained ties with former comrades, and read widely on military affairs. Visits from veterans and discussions of military affairs were common in the Clausewitz household, and the young Carl learned primarily by listening quietly to the stories of his father and his comrades.[11]

In the spring of 1792, shortly before his twelfth birthday, Clausewitz accompanied his father to Potsdam and joined the 34th Infantry Regiment.[12] There, he followed the path of his father and two brothers, enrolling as an officer cadet. All three Clausewitz brothers joined the army as children—driven to it by poverty, patriotism, and parental pressure—and ultimately rose to the general officer ranks. Carl's brother Wilhelm had joined the same unit five years earlier at the age of fourteen and had already risen to the rank of second lieutenant. Although Carl could rely on support from his brother, the shy, redheaded boy of slender build now faced a stark new world of rigorous Prussian discipline as he entered military service.[13]

By his own accounts, the twelve-year-old Clausewitz had mixed emotions about his new profession. Although he would have preferred to attend university or join the clergy, he understood these options were out of reach for him. Yet he had grown up hearing military stories from his father and family friends and was already a Prussian patriot, steeped in the glories of military life. Twenty-nine years later, as a major general, Clausewitz returned to the site of his old barracks and reflected upon those bygone days:

> I am well used to the fact that Potsdam always reminds me slightly of all kinds of serious and sad thoughts. It was always so. And it is natural enough, since I always feel strange and lonely here. I have returned to the house where I stayed with my father when he brought me for the start of twenty-nine years with the regiments. Not least, I have the highest feeling of gratitude for all the happiness which Fate has given me since that time.[14]

From an early age, Clausewitz developed a powerful, almost tragic, view of fate and destiny. Now, he put aside his youthful dreams and committed himself to mastering the military profession.[15]

Although the thought of a twelve-year-old boy joining the army is shocking by modern standards, it was not particularly unusual in late-eighteenth-century Prussia. As Frederick the Great quipped, the men of the Prussian officer corps were "snatched from their mothers' breasts" and forced into service of their country.[16] Indeed, so many children of minor nobles joined the military at such a young age that the Prussian military made specific allowances for these officer candidates. Rather than being directly commissioned, these aspiring officers were paired with a senior enlisted man who was charged with overseeing their military education and development. This unusual role has no direct analogue in modern militaries; the senior noncommissioned overseer functioned as a combination of drillmaster, first sergeant, and father figure. This system proved so effective at producing capable officers that it endured in various forms until the eve of World War II and has been credited, in part, with German military effectiveness.[17]

Clausewitz's First Taste of Battle: The Rhine Campaign

As the young Clausewitz was learning the rudiments of military life, war clouds were gathering. In their fervor to export the revolution, French forces launched a series of attacks into the Rhineland. These German states appealed to Prussia and Austria for assistance, and in early 1793, Clausewitz's regiment marched off to join what became known as the Rhine Campaigns. The campaigns consisted of a messy series of battles, sieges, and partisan warfare, with no clear victor. French revolutionary forces struck first—crossing the Rhine, terrorizing small communities, occupying Frankfurt for five weeks, and establishing their base of operations in Mainz. Local populations held mixed sympathies regarding the French invasion, but as word of alleged atrocities spread, Austrian and Prussian forces resolved to roll back the incursion into the German states. Marching under the slogan, "The Fatherland is in danger. The constitution, religion, property, tranquility—all are menaced by near ruin. The bloody projects of the French are unveiled," the Austro-Prussian armies rallied to the relief of Mainz.[18]

Clausewitz had not yet received his commission when he marched off to war in 1793. Despite his tender age and junior status, he served as flag bearer, leading the regiment from the front and earning incredulous stares from villagers as he marched in front of hundreds of battle-hardened Prussians veterans.[19] One can imagine the young Clausewitz's emotions —a mixture of fear, hope, curiosity, and enthusiasm—as he carried the flag at the head of his regiment. This emotional complexity—and war's mysterious nature—would remain with Clausewitz for the rest of his life. Indeed, he would dedicate his career to understanding its meaning—and there was much to learn in this first campaign.

When Clausewitz's regiment arrived at Mainz in late March, the French garrison of approximately 23,000 had already constructed extensive fortifications and prepared for a siege. The Austro-Prussian forces established their base of operations across the Rhine, and the siege of Mainz officially began on April 14, when they commenced bombardment of the French positions from across the river. In addition to the heavy cannon fire, both sides dispatched scouting parties, launched raids, and engaged skirmishers against the other's positions. During this fighting, Clausewitz was assigned to the regimental headquarters and tasked with inspecting hospitals and delivering messages between the command post and officers in the field. None of Clausewitz's writings from the period survive, but he appears to have performed his duties faithfully and earned the trust of his officers. His role as a messenger would have provided him with a unique, comprehensive view of the unfolding campaign as he moved across the battlefield—likely affording him a broader perspective than most of the front-line officers who remained tethered to their units.[20] Clausewitz would later write poetically about the shifting perspectives offered by different positions on the battlefield, suggesting that this early experience left a lasting impression on the young officer cadet.

The Austro-Prussian siege gradually gained momentum, and by June 18, 1793, they were positioned to subject Mainz to a massed bombardment. Firing from the heights above, the Austrian and Prussian forces quickly devastated the city below. As Goethe described the scene, "Every heart burned with sadness. Every moment one was filled with anxiety for one's dearest friends...For the gravely wounded one wished speedy release. And the dead one had no wish to recall life."[21] Filled with patriotic fervor, the young Clausewitz experienced conflicting emotions. As he later recalled, "I stayed while Mainz was being burned to the ground in the fire we had started...I added my childish shout to the triumphant cheers of the soldiers."[22]

The siege of Mainz became a formative moment in Clausewitz's early career. On July 20, while the siege was still ongoing, he was promoted to the rank of ensign. He was now a fully commissioned officer in the Prussian army. This was undoubtedly a proud moment for the young Clausewitz—a coming of age and a validation of his bravery and loyalty. This promotion relieved him of the duties of flag bearing and entitled him to ride on horseback with his fellow officers, don the officer's uniform, and issue commands. He was now also expected to fully share the dangers of officer service. Of the three ensigns in his unit, he was the only one not killed during the Rhine campaign, a grim testament to the front-line leadership demanded by the Prussian army.[23] Carl's brother Wilhelm further distinguished himself by rallying troops at an outpost threatened with being overrun. In an army that valued leadership and bravery, the Clausewitz brothers were quickly making a name for themselves as fearless junior officers. The king took note of the brothers' bravery and rewarded the Clausewitz sisters, Johanna and Charlotte, with the expectancy of rents from properties in Nuremberg and Marienborn, a significant financial windfall for the cash-strapped family.[24]

On July 22, two days after Clausewitz's promotion, Mainz surrendered. For four months, the city had been devastated by disease, famine, and cannon fire, and its will to resist had finally broken. As a condition to the French capitulation, the garrison was granted free passage out of the shattered city and was paroled on the promise not to bear arms for at least one year.[25] Such lenient terms may seem naïve to modern readers. However, these gracious terms were necessary to end the fighting before the political will of the Prussians and Austrians was exhausted. Prussian strategists recognized that they were resource constrained and questioned the costs and wisdom of keeping such a large portion of their brittle military deployed on a campaign far from home. The alliance with Austria was also becoming increasingly strained as national rivalries and petty grievances mounted. Viewed in this light, the relatively generous terms offered to the French garrison provide an excellent example of military considerations being subordinated to political ones—a fact not lost on Clausewitz, despite his youth and boyish enthusiasm for action.

The remainder of the Rhine Campaign proved frustrating for Clausewitz and Prussia. After the fall of Mainz, the Prussians seemed to lose both drive and purpose. They failed to pursue the retreating French forces across the Rhine, missing an opportunity to achieve a decisive victory over a broken foe. Clausewitz later commented in *On War*, "Prussia had nothing to defend or conquer in Alsace. Her march in 1792...had been made in the spirit of chivalry, but since as things turned out that operation promised little more, she pursued the war without enthusiasm."[26] An uprising among Prussia's Polish population further sapped political will, forcing many troops to be withdrawn toward the homeland. Any remaining hope for a decisive victory soon faded. These setbacks taught Clausewitz valuable lessons about the importance of political will, the difficulty of sustaining large forces in the field, and the primacy of defensive forms of warfare.[27]

Ultimately, Prussia exited the war in April 1795 by signing the Peace of Basel.[28] Austria continued to fight an increasingly resurgent France under Napoleon Bonaparte until she was forced to accept the Peace of Campo Formio in October 1797.[29] Through dithering and infighting, the Prussians and Austrians squandered their opportunity to deliver a decisive defeat to France and inadvertently contributed to the rise of their greatest nemesis: Napoleon. Clausewitz lamented these failures, yet he was in no position to influence events. He was promoted to second lieutenant in March 1795 but saw no further action before the conclusion of hostilities. His regiment was posted to the bucolic garrison town of Neu-Ruppin where, bereft of both action and educational opportunities, Clausewitz quickly began to flounder.

Neu-Ruppin: A Frustrated Autodidact Emerges
Clausewitz had much to be proud of following his first campaign. He had earned multiple promotions, won the respect of his fellow officers, demonstrated his bravery, and gained valuable lessons about the nature of war. Yet, for the enterprising junior officer, this was not enough. He had developed a taste for action and wanted more, but his nation was at peace and his unit was now stationed in the pastoral town of Neu-Ruppin. He detested the slow pace of garrison duty and soon developed a particular aversion to drill and ceremony. He regarded formal maneuvers as a waste of time, bearing little connection to the actual combat he had just experienced.[30]

At the age of fifteen, rather than enjoy the serenity of his surroundings, the deeply ambitious Clausewitz embarked on a period of intense self-study and personal improvement.[31] As fate would have it, the barracks at Neu-Ruppin stood near the residence of Frederick the Great's brother, Prince Henry, whose royal home included a vast library that was open to the regiment's officers.[32] There, Clausewitz read broadly from Enlightenment philosophers and explored more contemporary texts from German thinkers such as Friedrich Schiller, Wolfgang von Goethe, and Friedrich Hölderlin.[33] In addition to philosophical works, Clausewitz read extensively in military history and strategy. For a low-ranking subaltern confined to garrison duty, such self-directed study was highly unusual and reflected a remarkable determination to improve his education and deepen his understanding of military affairs.

Clausewitz appreciated his self-study but he recognized that it was unstructured and amateurish, and he ultimately found it unsatisfying. This frustration with military routine and solitary academic endeavors seem to have exacerbated his own shyness and likely marked him as an intellectually gifted but socially withdrawn young man.[34] He was tempted to leave the army, but with no financial support for further study and no other work experience, he had few viable alternatives. What Clausewitz needed was mentorship, structure, and an opportunity to distinguish himself, but he would have to wait six years before taking the next significant step in his career.

An Education and a Mentor

Clausewitz's malaise came to an end in 1801, when he applied and was accepted into the Institute in the Military Sciences for Young Infantry and Cavalry Officers in Berlin.[35] This institution, now known as the Prussian War College, had been recently transformed from a poorly regarded finishing school for officers into a rigorous institution for the study of war. These reforms were due largely to the efforts of Gerhard von Scharnhorst.[36] A formidable intellect in his own right, Scharnhorst believed that military education and the mentorship of junior officers had been woefully neglected. To address this, he designed a curriculum that aimed to cultivate the minds of young officers by blending theoretical study with practical application. At the same time, he provided them with a liberal education and required them to solve practical military problems as if they were a commanding officer on campaign.[37]

Such a course of study had an innate appeal to Clausewitz, who sought recognition not only as a scholar and thinker but also as a military leader and man of action. As he traveled to Berlin to attend school, Clausewitz recognized this as a critical moment in his life because he had reached the limits of his informal self-study.[38] While Clausewitz was eager to engage his intellect, he was concerned about the social requirements of life in Berlin. Relative to the elite of Prussian society, he was penniless, shy, and perceived as provincial. He could remain unnoticed in Neu-Ruppin, but an assignment to the war college required him to move to Berlin and participate in elite social circles. For the introverted and finan-cially strained young officer, this forced acculturation was intimidating; however, it would prove just as important to his later success as his academic studies. Indeed, during this period, he met the king, the royal court, a cadre of Prussian military reformers, his two greatest mentors — Gerhard von Scharnhorst and August Neidhardt von Gneisenau—and his future wife, Marie von Brühl.[39]

At the war college, Clausewitz began a three-year course of instruction designed to prepare him for higher command and staff work.[40] While this was the educational opportunity that Clausewitz had long wanted, he seemed to have suffered a crisis of confidence during his first weeks at the war college. He worried that he would never succeed and seriously considered withdrawing from the school and returning to his regiment.[41] As his wife Marie would later admit, "In the beginning, it was very hard for him to follow the lectures because of his lack of essential knowledge. He was close to despair and probably would have given up on the hard endeavor."[42] Fortunately, Clausewitz's innate talents were recognized by a prominent member of the faculty, Colonel Gerhard von Scharnhorst, who took him under his wing and "encouraged him with his unique kindness and benevolence and at the same time through his enlightened teaching woke up and developed all the seeds of his mental resources."[43]

Buoyed by Scharnhorst's confidence, Clausewitz renewed his commitment to his studies and quickly flourished in the rigorous academic environment. At last, he was able to benefit from the educational opportunities that had so long been unavailable to him. He understood that his time at the war college was a pivotal moment in his development as an officer. Success here would affirm that he had become the scholar and soldier that he had long aspired to be. As a poor boy of small stature and shy demeanor, Clausewitz was considered an underdog, but he dedicated himself to achieving excellence in this new endeavor. These efforts paid off, and by March 1802, Clausewitz had distinguished himself as one of the top students at the war college. One of his class papers was recognized for its exceptional quality and sent to the king, an extraordinary achievement for a student who had nearly withdrawn in defeat.[44] The king remarked on this paper, suggesting that Clausewitz was destined for high rank and greater responsibility.[45]

As part of the war college curriculum, Clausewitz read philosophical works from a wide range of authors including Enlightenment thinkers such as Dupont, Voltaire, and Montesquieu.[46] Montesquieu quickly became one of Clausewitz's favorite writers because he appreciated the subtle and layered meanings within the classic *Spirit of the Laws*.[47] He wrote that these "precise, aphoristic chapters" kindled his interest "by what they suggested as much as by what they expressed."[48] To this end, Clausewitz read and reread these Enlightenment authors and attempted to model his own prose on their sophisticated and occasionally obscure style.[49] From these scholars, he learned the importance of thinking systematically about a problem and using reason and evidence to fully develop a concept. While he would ultimately reject many of the progressive assumptions of the Enlightenment, Clausewitz took these ideas seriously and refined his own assumptions in the process.[50]

During this period, Clausewitz was also exposed to Machiavelli and other classical realist thinkers. Machiavelli also became one of Clausewitz's most significant influences because Clausewitz valued his insight that war and the threat of violence underpinned all political life.[51] In contrast to the Enlightenment writers' optimistic spirit, Machiavelli's stark realism appealed to Clausewitz, who remarked, "No reading is more necessary than the writing of Machiavelli. Those who pretend to be revolted by his principles are nothing but dandies who take humanist airs."[52] Machiavelli's willingness to describe war and politics as they actually were, rather than as a romanticized ideal, deeply impressed Clausewitz.[53] Clausewitz's later writings on the political purposes of warfare and the need for statesmen to act rationally and amorally were clearly inspired by the classical realist tradition in general and Machiavelli in particular.[54]

As Clausewitz matured, he would employ elements from both philosophical traditions. He attempted to apply a rationalist, Enlightenment approach grounded in reason and evidence, while also viewing the world through the tragic lens of realism, recognizing the world as flawed and resistant to logical explanation.[55] This internal tension may give the impression of inconsistency, but Clausewitz belonged fully to neither the Enlightenment nor the anti-enlightenment camps.[56] Rather, from these early days, he sought to synthesize the best of both approaches to make sense of war—a complex phenomenon that , in his view, was intelligible though not entirely rational.

In addition to these classroom experiences, Clausewitz also began to cultivate personal relationships with key members of the Prussian military elite. Of these, none was more important than his relationship with Scharnhorst.[57] Scharnhorst encouraged the reserved Clausewitz to remain at the war college and proved to be an ideal mentor. He possessed a powerful intellect, a fierce patriotism, and a passion for military innovation and reform. This relationship also fulfilled Clausewitz's need for a male role model, which had been absent since his father enrolled him in the military at the age of twelve. Scharnhorst was older, more mature, and had already achieved fame and success of his own. A starstruck Clausewitz was all too willing to be molded by a man he later described as, "father and mother and friend of my soul."[58] This relationship was of critical importance and should not be underestimated. Clausewitz had gained an inspiring mentor who would guide him through a period of intellectual and personal discovery and provide him with a firsthand view of the Prussian military reform movement. While many of Scharnhorst's efforts during this period were frustrated by institutional malaise, he was already an impressive soldier and military leader, an intellectual innovator, and a mentor to a generation of like-minded officers, and he enjoyed direct access to the elite strata of Prussian society. In 1809, he was appointed head of the Prussian army and began a sweeping transformation of the military system.

Scharnhorst was particularly impressed by the young Clausewitz's academic potential. He noted that Clausewitz and his friend, Carl Ludwig Heinrich von Tiedemann, were the two best students in class and that "they were distinguished in talent, judgment and industry and by their altogether exceptional knowledge." Scharnhorst also noted: "The character of Lieutenant von Clausewitz is distinguished by being modest and pleasant. Moreover, he possesses a thorough knowledge of mathematics and military scholarship."[59]

To further stimulate Clausewitz's intellectual talents, Scharnhorst invited him to join the prestigious *Militärische Gesellschaft* (Military Society) an elite, semi-official association of leading Prussian military scholars and reformers.[60] The members gathered to study and discuss key military topics and were actively involved in efforts to reform the traditionally conservative Prussian army. In this intellectually stimulating environment, Clausewitz continued to develop his ideas and analytical skills.[61]

Clausewitz's 1802 fitness report reflected both his newfound purpose as well as the influence of Scharnhorst's mentorship:

His conduct is good; he is a good officer who seeks to acquire knowledge. Is presently in Berlin to attend the military courses, where he is supposed to be very industrious and according to the judgement of Colonel von Scharnhorst one of the brightest minds.[62]

For the first time in his career, Clausewitz had found genuine purpose and fulfilment through the pursuit of knowledge. This awakened a deep intellectual ambition that would shape the course of his life.

The Love of a Lifetime

In addition to his academic successes, Clausewitz also found love during his time as a war college student. This was unexpected as he was shy around women, and it was unusual for junior officers without financial means to court or marry during this period. Despite these limitations, Clausewitz met the aristocratic Marie von Brühl at a dinner in 1803. Marie, a lady-in-waiting to the dowager queen, was a witty and sophisticated courtier. She was immediately attracted to the young officer, but he was too shy to speak with her. He initially described her as a "bright woman" but seemed to regard her as a friend rather than a romantic prospect.[63]

In spite of this inauspicious beginning, the two kept crossing paths in the months ahead, and as they took note of each other's intellectual gifts, they began to develop a powerful bond.[64] Their relationship evolved slowly at first, restricted by their infrequent meetings, their obligations to the military and the royal court, and the social mores of the period. While they admired each other's intellect, theirs was not, in many respects, a logical match. He was an unknown junior officer with limited means and lower social standing, while she descended from a prominent family and was part of the inner circle of the royal court. Because of her higher social standing, Marie's mother disapproved of the match, encouraged her daughter to seek a wealthier and more socially prominent husband, and ultimately blocked their marriage until 1810.[65]

Nevertheless, Marie persisted in her pursuit of the young officer. This revealed not only their shared affection but also Marie's strength of character. Her family had experienced serious financial strain since the death of her father in 1802, and she would have known that Clausewitz had limited financial prospects. Given her royal connections, she could have expected a more advantageous match. But to her mother's frustration, Marie refused to marry a series of older and wealthier men introduced to her. Ultimately, the two became engaged and officially became a couple in 1806. They married four years later, after Clausewitz had been promoted to major and received official permission from the king.[66] While unconventional, their love was genuine, and they would become close intellectual companions whose legacies would become intimately intertwined.

From Graduation to Early Success

Thanks to his hard work, natural intelligence, military aptitude, and initiative, the young Clausewitz graduated from the war college at the top of his class in the spring of 1804.[67] This was a major accomplishment for Clausewitz as it partially satisfied the intellectual hunger he had felt since his early years and marked him as a future leader in the Prussian military. After graduation, he leveraged his academic credentials and relationship with Scharnhorst to serve as an aide-de-camp for Prince August.[68] This was a highly prestigious appointment for the young Clausewitz and signaled his rapid ascent within the esteemed Prussian military.

Building on this success, Clausewitz published his first article on military affairs in 1805.[69] This article was a methodological critique of General Heinrich Dietrich von Bülow's latest work on military strategy. With Scharnhorst's support and encouragement, Clausewitz published his first article, an achievement that brought significant pride to both men. In this article, Clausewitz began developing some of the theories he would continue to refine over the next quarter century. He criticized Bülow for failing to make his assumptions and hypotheses explicit. He argued that Bülow's work was unrealistic because it ignored the physical and emotional aspects of battle, and he emphasized that any theory of war must account for the political aims of the belligerents.[70]

In an undated essay, "The Theory of the Art of War Today," which appears to have been written during this period, Clausewitz further expanded on his frustration with the current state of military theory. He was similarly dismayed by the lack of theoretical rigor and by how disconnected the theorists' perspectives were from the realities of war. He gave voice to his frustrations, noting, "Whatever great achievements have occurred in the history of war, they were not due to books."[71] These were important insights for the young Clausewitz. His methodological desire to clearly state assumptions and theses informed his dialectic approach, while his substantive criticisms laid the foundation for his most influential theories: friction and the idea of war as an extension of politics.

All seemed to be going well for the young Clausewitz. He was flourishing as a thinker and military officer; he was making key connections within the Prussian elite and enjoying his first successes as an author. Unfortunately, the venerated Prussian military system was about to experience an unexpected upheaval that would have far-ranging consequences for the aspiring general and his nation.

Notes

1. Paret, *Clausewitz in His Time*, 5.
2. Bassford, *Clausewitz in English*, 27; and Waldman, *War, Clausewitz, and the Trinity*, 109.
3. Clausewitz, *On War* (Howard and Paret), II.3, 148.
4. Strachan, *Clausewitz's On War*, 33.
5. Stoker, *Clausewitz: His Life and Work*, 6.
6. Bellinger, *Marie von Clausewitz*, 28.
7. Bassford, *Clausewitz in English*, 10.
8. Paret, *Clausewitz and the State*, 18–19.
9. Parkinson, *Clausewitz: A Biography*, 21–22 and 35.
10. Paret, *Clausewitz and the State*, 18–19.
11. Paret, 13–15; and Parkinson, *Clausewitz*, 21–22.
12. The date here is somewhat uncertain. It appears possible that Carl's father deliberately altered his son's birthday so that he could join the unit sooner, gain seniority, and be included in the forthcoming campaign; Bellinger, *Marie von Clausewitz*, 28 and Paret, *Clausewitz and the State*, 19.
13. Parkinson, *Clausewitz: A Biography*, 15–23; Stoker, *Clausewitz: His Life and Work*, 4; and Strachan, *Clausewitz's On War*, 37.
14. Parkinson, *Clausewitz: A Biography*, 19.
15. Lebow, *The Tragic Vision of Politics*, 168–215.
16. Parkinson, *Clausewitz*, 34; and Stoker, *Clausewitz: His Life and Work*, 4.
17. Muth, *Command Culture*; and Parkinson, *Clausewitz: A Biography*, 15–22.
18. Parkinson, *Clausewitz: A Biography*, 23.
19. Parkinson, 23.
20. Parkinson, 23–24. For Goethe's account, see Goethe, *Miscellaneous Travels of J.W. Goethe*, 249–288.
21. Stoker, *Clausewitz: His Life and Work*, 12–15.
22. Bellinger, *Marie von Clausewitz*, 32.
23. Stoker, *Clausewitz: His Life and Work*, 22.
24. Unfortunately for the cash-strapped Clausewitz family, the sisters found it nearly impossible to obtain the funds they were owed as part of this reward; Parkinson, *Clausewitz: A Biography*, 25–26.
25. Parkinson, 26.

26. Clausewitz, *On War* (Howard and Paret), VIII.9, 631.
27. Paret, *Clausewitz and the State*, 23–30.
28. Bellinger, *Marie von Clausewitz*, 33; and Lebow, *The Tragic Vision of Politics*, 171.
29. Nester, *Napoleon and the Art of Diplomacy*, 33; and Parkinson, *Clausewitz*, 29.
30. Paret, *Clausewitz and the State*, 45.
31. Aron, *Clausewitz: Philosopher of War*, 231.
32. Howard, *Clausewitz*, 6; and Stoker, *Clausewitz: His Life and Work*, 25.
33. Paret, *Clausewitz and the State*, 41.
34. Parkinson, *Clausewitz: A Biography*, 29; and Strachan, *Clausewitz's On War*, 38–39.
35. Stoker, *Clausewitz: His Life and Work*, 28.
36. White, *The Enlightened Soldier*, 39, 49,79–80, 87–88, and 112.
37. This approach resembled the method of critical analysis that Clausewitz employed—though with varying degrees of clarity and success—in *On War*; Clausewitz, *On War* (Howard and Paret), II.5, 156–169. On the curriculum of the War College, see Paret, *Clausewitz in His Time*, 18–22; and Strachan, *Clausewitz's On War*, 39.
38. Strachan, *Clausewitz's On War*, 39.
39. Bellinger, *Marie von Clausewitz*, 13.
40. Paret, *Clausewitz in His Time*, 18; and Stoker, *Clausewitz: His Life and Work*, 30.
41. Stoker, *Clausewitz: His Life and Work*, 30.
42. Bellinger, *Marie von Clausewitz*, 43; and White, *The Enlightened Soldier*, 101.
43. Bellinger; and Paret, *Clausewitz and the State*, 74–75.
44. Paret, *Clausewitz in His Time*, 26; and White, *The Enlightened Soldier*, 100.
45. Scharnhorst intended for the war college to serve as a recruitment pipeline for the General Staff and appears to have sent papers to the king as a means of grooming his star pupils; Craig, *The Politics of the Prussian Army, 1640–1945*, 45.
46. Paret, *Clausewitz and the State*, 4; and Parkinson, *Clausewitz: A Biography*, 35.
47. Aron, *Clausewitz*, 60.
48. Strachan, *Clausewitz's On War*, 89.
49. Aron, *Clausewitz*, 215 and 232.
50. Waldman, *War, Clausewitz, and the Trinity*, 20–21 and 26–27.
51. Heuser, *Reading Clausewitz*, 7, 26, 44, and 80.

52. Aron, *Clausewitz*, 20; Parkinson, *Clausewitz: A Biography*, 35; Strachan, *Clausewitz's On War*, 88–89 and 163; and Waldman, *War, Clausewitz, and the Trinity*, 8–9, 65, 104, 107–109, 124–125, and 129–130.

53. Clausewitz was so inspired by Machiavelli's writings, that in 1809 he corresponded with the idealist philosopher Johann Gottlieb Fichte to discuss his recently published book on Machiavelli; Heuser, *Reading Clausewitz*, 7; and Lebow, *The Tragic Vision of Politics*, 186.

54. On Clausewitz's intellectual connection to Machiavelli, see Paret, *Clausewitz and the State*, 81. Interestingly, the Greek philosopher Plato was the first known writer to relate war to politics, though there is no evidence that Clausewitz took inspiration from this source; Coker, *Barbarous Philosophers*, 86.

55. Waldman, *War, Clausewitz, and the Trinity*, 26–27.

56. Lebow, *The Tragic Vision of Politics*, 178.

57. Paret, *The Cognitive Challenge of War*, 79–82.

58. Parkinson, *Clausewitz: A Biography*, 32; and Strachan, *Clausewitz's On War*, 40.

59. Much to Clausewitz's sorrow, Tiedemann died during the Russian campaign in service to the tsar; Stoker, *Clausewitz: His Life and Work*, 132.

60. For an overview of the Military Society, see White, *The Enlightened Soldier*, 30, 37–42, 49–50, 183, and 187.

61. Echevarria, *Clausewitz and Contemporary War*, 43–44; Paret, *Clausewitz and the State*, 66–67; and Paret, *The Cognitive Challenge of War*, 80.

62. Paret, *Clausewitz and the State*, 55.

63. Bellinger, *Marie von Clausewitz*, 13 and 51–52.

64. Bellinger, 13, 38, 41, and 47.

65. Bellinger, 41–47 and 60–61; Paret, *Clausewitz and the State*, 99; and Strachan, *Clausewitz's On War*, 42.

66. Bellinger, *Marie von Clausewitz*, 47–50, 63, and 104.

67. Echevarria, *Clausewitz and Contemporary War*, 44; and Stoker, *Clausewitz: His Life and Work*, 31.

68. Paret, *Clausewitz and the State*, 75.

69. Clausewitz had written numerous essays prior to this period, including one in 1803 that compared Napoleonic France to the Roman Empire. Despite his literary productivity, he was consistently reluctant to publish his work; Heuser, *Reading Clausewitz*, 2; and Stoker, *Clausewitz: His Life and Work*, 32–33.

70. Paret, *The Cognitive Challenge of War*, 113–114.

71. Paret, 116–117.

CHAPTER 3

WAR AND DEFEAT: CLAUSEWITZ'S RESPONSE TO FAILURE

Jena and Auerstädt and the Disaster of 1806
In many ways, the 1806 war was a clash of two profoundly different military systems: the smaller, parochial, and traditional army of Prussia and the larger, more modern revolutionary army of France.[1] On nearly every level, the Prussians were clearly outmatched. Diplomatically, they had failed to support previous anti-French coalitions when their assistance might have tipped the balance, and they then declared war before their Russian allies were ready to come to their aid. Operationally, the Prussians suffered from an outdated system of command and control which rendered their disorganized forces cumbersome and slow during campaigns. The Prussians were also inferior on the tactical level, relying on formations and doctrine that had changed little since the days of Frederick the Great.[2]

While these flaws may have been clear in retrospect, the Prussian army entered the 1806 campaign, unaware of how overmatched it truly was. Even the typically sober-minded Clausewitz failed to grasp the full gravity of the situation. On September 16, 1806, he wrote to Marie in an almost manic tone: "War alone can enable me to achieve happiness. In whatever way I wanted to connect my life to the world as a whole, my path led me through a major battlefield." He wrote to Marie again on September 29, equally optimistic: "If I draw a conclusion from all the observations I have made, it seems to me probable that we shall be the victor in the next great battle."[3]

On paper, Clausewitz had reason to be optimistic. The Prussians had Napoleon's forces trapped between their armies. They planned to win a decisive victory by deploying quickly from their bases and striking from two directions simultaneously while the French forces were vulnerable. The Prussians hoped that they could force a decisive campaign, save Prussia, and bring an end to Napoleon's domination of Europe. Unfortunately, the Prussians were slow to execute their movements, and Napoleon did not passively accommodate their plan.[4] Instead, the French emperor relied on superior organization, experience, and the concentration of his forces to achieve victory.[5] Advancing from his interior lines, he moved at a speed that, according to Clausewitz, was, "unheard of in the history of warfare."[6]

Using his speed and interior position, Napoleon first defeated the Prussian force to the south at Jena and then shifted his army north and won another victory at Auerstädt. In these two battles, the Prussians suffered 20,000 killed or wounded and 18,000 captured, out of a total force of 103,000. In addition to this 37 percent casualty rate, the Prussian artillery corps lost 235 of their 298 field guns.[7] In a single day, Prussia suffered the greatest defeat in its history. As Clausewitz would later write about the impact of defeat on the morale of an army:

> Those who have never been through a serious defeat will naturally find it hard to form a vivid and thus altogether true picture of it... When one is losing, the first thing that strikes one's imagination...is the melting away of numbers. This is followed by a loss of ground...

Next comes the break-up of the original line of battle, the confusion
of units, and the dangers inherent in retreat...Then comes the
retreat itself, usually begun in darkness, or at any rate continued
through the night. Once that begins, you have to leave stragglers
and a mass of exhausted men behind...The feeling of having been
defeated, which on the field of battle had struck only the senior
officers, now runs through the ranks down to the very privates...
Worse still is the growing loss of confidence in the high command...
the sense of being beaten is not a mere nightmare that may pass:
it has become a palpable fact that the enemy is stronger...one is
harshly and inexorably confronted by the terrible truth.[8]

Clearly, Clausewitz was deeply affected by the Prussian defeat and would
carry this pain with him for the remainder of his life.

On a broader level, these defeats signified a fundamental shift in the
conduct of war. The old paradigm of Frederick the Great, defined by
small professional armies, was shattered by Napoleon.[9] In these twin
battles, the mass, patriotic fervor, and superior battlefield performance
of the French forces proved decisive. Clausewitz noted that "Prussian
troops using Frederician tactics with courage and determination [were
expected] to overcome anything that had emerged from the unsoldierly
Revolution," yet this was not to be.[10] Instead, the Prussian defeat signaled
the beginning of the end for this proud but antiquated military system.[11]

Clausewitz Captured
In the aftermath of these twin defeats, the Prussian army was in a state
of chaos. The army attempted to retreat in an orderly manner toward
Berlin in the hope of reorganizing and staging a defense of the capital,
but these plans quickly disintegrated when the death of General Charles
William Ferdinand Brunswick left the Prussians in a state of leaderless
confusion. Napoleon remorselessly pursued his defeated foes, capturing
many Prussian stragglers and forcing countless others to abandon their
equipment or desert their units.

As the Prussian forces crumbled, Clausewitz continued to serve honorably. Along with Prince August, he fled with a group of soldiers attempting to escape and rejoin the Prussian main body. In the face of a broken chain of command, Clausewitz took control of these men and did his best to maintain order. After forming infantry squares and fighting off repeated attacks by French cavalry, his small band of stragglers was forced to surrender as they came under artillery fire and their escape route was blocked. With no option to escape or resist, Clausewitz and Prince August became two of over 18,000 Prussians captured at the twin battles of Jena and Auerstädt.[12] Over the next month, an additional 122,000 Prussians were captured, a truly staggering total that effectively eliminated their ability to resist.[13]

Clausewitz was now a prisoner but was allowed to remain with Prince August. August was taken to Napoleon's tent for interrogation, and the then-unknown Clausewitz was allowed to wait outside. August and Clausewitz were now prisoners of the French but were treated with respect and care due to their rank and social standing. Initially, they were allowed to remain free and on Prussian soil, provided they swore not to bear arms until they were officially paroled. Prince August and Clausewitz settled into a relatively luxurious captivity in the Prussian village of Neu-Ruppin, the same garrison town where, as a younger man, he had first begun his self-study of philosophy and the military arts. Here, they waited and read the increasingly bleak news about the fate of their beloved nation.[14]

After the stunning victories at Jena and Auerstädt, Napoleon's forces aggressively pursued the remnants of the Prussian army. The Prussians retreated toward Berlin but were unable to properly reorganize their forces and mount an effective defense.[15] Berlin fell a few weeks later after token resistance, and the Prussians were at the mercy of Napoleon's forces. The effectiveness of the pursuit after the battlefield victories at Jena and Auerstädt left a strong impression on Clausewitz. He recognized that the exploitation of initial successes was key to achieving more permanent political objectives. He later incorporated this powerful lesson into his understanding of war, noting that "victory consists not only in the occupation of the battlefield, but in the destruction of the enemy's physical and psychic forces, which is usually not attained until the enemy is pursued after a victorious battle."[16] Such a vigorous pursuit forces a foe into "continuous uninterrupted flight" and will break their will to resist because "nothing is more repugnant to a soldier than hearing the enemy's guns yet again just as he is settling down to rest after a strenuous march. This sensation, repeated day after day, can lead to absolute panic."[17]

Prussia was disarmed and its capital was occupied. In the aftermath of his complete victory, Napoleon was the master of the European continent and could unilaterally decide the fate of Prussia. In response to the British blockade of European ports, the French emperor announced the Berlin Decree on November 21, 1806.[18] Here, he declared that all the nations of Europe were part of the Continental System and that they must halt all trade with Great Britain. While this declaration ultimately sowed the seeds for Napoleon's destruction, it emphasized how he had come to dominate the landmass of continental Europe.[19]

Clausewitz's writings from this period reveal a mixture of sadness, anger, and helplessness as he struggled to comprehend a seemingly endless series of military and political humiliations. While he could not fight until he was paroled, Clausewitz poured his frustrations into an attempt to document, analyze, and remedy the Prussian failures. His first efforts were a series of notes about the tactical lessons he had recently witnessed. He praised the efforts of his troops to fend off French cavalry and noted that so long as they maintained their order and discipline, they could be expected to fend off repeated mounted charges. While this can be seen as a defense of his actions prior to his capture, he was contributing to a contemporary debate about the ability of cavalry to break infantry squares by arguing in favor of well-disciplined foot soldiers.[20]

Moving on from this rather narrow point, Clausewitz composed three articles during his time in Neu-Ruppin, which would be published a year later in the military journal *Minerva*. These three pieces were much more ambitious and provocative as Clausewitz was deeply critical of the Prussian military. He argued that the retreat from Auerstädt did not need to lead to disaster but was the result of poor organization and leadership as well as the ability of the French forces to regroup and reengage. While the Prussians were rendered helpless, the French suffered significant losses yet were able to press their advantage. According to Clausewitz, the key difference was one of leadership and fighting spirit. Prussian generals moved slowly and missed multiple opportunities to improve their situation. These repeated failures exhausted the fighting spirit of their men and thus led to their ruin. This was a very bold claim to make as it struck directly at the heart of the Prussian military system by challenging the institution's prized qualities of discipline and leadership. Although these articles had a few minor factual errors, they correctly summarized many of the Prussian failures. More importantly, they are a true reflection of Clausewitz's experiences during the campaign and of his desire to change the system.[21] Years later, while composing *On War*, he would incorporate this insight about the need to act quickly and boldly as the two main components of military genius, a clear indication of how deeply this military catastrophe influenced his thinking.

In early December 1806, Clausewitz received word that Prince August was being sent to captivity in France. As the prince's aide-de-camp, Clausewitz was expected to accompany him into captivity and to depart Prussia by the end of the month. This move interrupted his writing and courtship of Marie and deepened his sense of loss and ennui. Clausewitz and the prince departed on December 30 and arrived in Nancy on January 18, 1807, before being sent to Soissons in March. While Clausewitz admired the scenic countryside and marveled at the Reims cathedral, his journey through France kindled a deep hatred of Napoleon. Almost as soon as he arrived at Soissons, he began to write an aspirational campaign plan for an Austrian invasion of France. Nothing came of this document, but it clearly indicates how frustrated Clausewitz was with his new surroundings and how he attempted to calm his frayed nerves by engaging his mind on a military task, however fanciful it may have been.[22]

As high-status prisoners, Clausewitz and the prince enjoyed a life of relative creature comfort. Because he was an officer in the service of the prince, Clausewitz was allowed to travel around France on his own recognizance and could receive mail, read, and generally live a life of introspection and leisure. He traveled with August to Paris, where, at Marie's insistence, he visited museums, churches, art galleries, and salons. However, Clausewitz was not in the mood to enjoy himself and remarked that people only cared about masterpieces such as Raphael because of their high price, not their artistic merits.[23] Despite the fact that he was an enemy noble in revolutionary France, Prince August was treated as a celebrity by his French hosts. Much to Clausewitz's dismay, the young prince was particularly fond of bordellos, salons, and the many sensual pleasures that were made available to him. The prince quickly developed a reputation as an extroverted *bon vivant*, while Clausewitz, in contrast, withdrew into a quiet depression. His letters from this period indicate a feeling of being trapped—literally by his French captors and metaphorically by his association with the indulgent Prussian noble.[24] He wrote to Marie that he felt like a general who had been lured into enemy territory only to discover that "a hostile army was at his rear."[25]

Despite his relatively easy life, Clausewitz chafed at being in captivity and began to plan for the military rebirth of his beloved Prussia. As was his nature, Clausewitz turned to his pen. He co-authored a series of memos with August documenting the causes of the Prussian defeats. Clausewitz appears to have done the majority of the work on these reports, using the prince's name to lend his ideas official legitimacy. He identified two critical flaws that led to the disaster of 1806: the slowness of Prussian movements and the poor communication and coordination between various branches of the army. To remedy these deficiencies, Clausewitz proposed universal conscription; an end to the employment of mercenaries and to the practice of impressment; greater inclusion of

officers from the bourgeoisie; and the abolition of harsh discipline. Each
of these overlapping reforms would improve the army's fighting spirit
and initiative of the army by bringing in new talent from a broader cross-
section of society, increasing meritocracy within the force, harnessing
the power of nationalism, and incentivizing creativity and ingenuity.
Following the French revolutionaries' example, Clausewitz wrote:

> It is every citizen's duty to defend the state...the State acquires
> a great military power when this principle is raised to a law
> and every soldier made eligible to the higher ranks...Nothing
> runs counter to this so much as corporal punishment and the
> unfortunately still widespread custom of forcing men to serve in
> the army. Both must be abolished if military service is to agree
> with the principles according to which the soldier should act.[26]

Even though he had written these memos using the prince's name,
Clausewitz understood that they contained potentially radical and disruptive ideas. Rather than present them directly to the newly established
Military Reform Commission, Clausewitz—naturally reserved—forwarded
these suggestions to Scharnhorst, who received them with great interest.
Scharnhorst had been thinking along similar lines and wrote Clausewitz
a heartfelt letter thanking him for the memo and requesting a reunion
after his release.[27] The importance of these memos should not be underestimated. While in captivity, Clausewitz had directed his frustrations
into a series of documents that correctly diagnosed the problems endemic
to the Prussian military, proposed solutions that would become the heart
of the Prussian reform movement, and established himself as a critical
and disruptive thinker who had the support of the prince.

Ultimately, Clausewitz's period of captivity would end following
Napoleon's victory over the Russians at the Battle of Friedland in June
1807.[28] In the aftermath of this decisive defeat, Russia was forced to sue
for peace rather than face the prospect of losing significant territory
to the advancing French army. The resulting Treaties of Tilsit (July 7
and 9) helped preserve Russian power but were a further humiliation

to a helpless Prussia. Under this extreme duress, Prussia was forced to cede approximately half of its territory, pay crushing reparations, become an ally of France, and reduce its army to a mere 42,000 men.[29] The Prussian sovereign, Frederick Wilhelm III, was allowed to remain on the throne, but he was a deeply flawed leader whom Napoleon saw as little more than a hapless puppet.[30] After this defeat, Wilhelm became increasingly suspicious, petty, narrow-minded, insecure, and fearful of provoking the ire of Napoleon. Yet, despite these shortcomings, the king secretly remained a fierce patriot who dreamed of a reborn and revitalized Prussian army.[31]

Intellectual Growth from Military Humiliation

For Clausewitz, the defeats of 1806 and 1807 provided a lifetime of motivation. In the months following the Treaty of Tilsit, Prussia suffered some of the worst disasters in its history. Epidemics of cholera, typhoid, and dysentery ravaged the population. As a result of the Continental System, food supplies dropped rapidly, and prices soared. Millions went hungry, and infant mortality in Berlin reached as high as 75% for several months. As the population faced this devastation, land prices dropped and businesses failed.[32] Clausewitz lamented:

> The bankruptcies here are endless...what was achieved in this sandy waste throughout centuries in the way of prosperity, culture, and trade, will now be destroyed in perhaps a decade.[33]

These deprivations deepened his nationalism and hatred of the French, forcing deeper thinking about the need for military reform.[34]

While he had always had an inquisitive mind, nothing motivated Clausewitz as much as understanding this calamity and ensuring that his beloved Prussia would never again be at the mercy of a foreign power.[35] He described his feelings at length in an 1813 article:

> In the unfortunate days of Jena and Auerstädt, the Prussian army lost its glory; in the retreat it fell apart. Its fortresses were given

up, the state was conquered, and after four weeks of fighting little
was left of either state or army...[the Treaty of Tilsit] completed
the misery...Within a year, Prussia's glittering military state, a
joy to all lovers of soldiers and war, had disappeared. Admiration
was replaced by reproach and censure, homage by humiliation.
An oppressive sadness weighed on the army's morale. Finding
confidence in the past was not possible; nor was hope in the future.
Even that ultimate source for regaining courage, trust in particular
leaders, was absent, because in the brief war no one had achieved
prominence, and the few who had distinguished themselves, were
divided among factions holding different opinions....In view of the
army's depressed spirit, the state's weakened economy, its financial
ruin, imperious interference from without and discouragement
within, which blunted every energetic measure, it was hard to
reach the goal we set out for ourselves...[the goal was to] Renew
and encourage the army, lift its spirits, eradicate old flaws, and as
it was trained and built up to the strength allowed, lay the basis
for a new, larger force that would be ready to spring into action
at some future decisive moment.[36]

Clearly, Clausewitz felt the impact of these defeats personally and
was motivated by an intense combination of intellectual curiosity and
patriotism. More than 20 years after these battles (the exact date is
unknown), he would return to the subject of Prussian defeat for a third
time with his *Observations on Prussia in Her Great Catastrophe*. This was a
work of considerable sophistication and breadth, which included analysis
of political and cultural factors that led to the military disaster. He noted
candidly that the Prussian "machinery of government was desiccated,
decrepit, and entirely unsuited to the times."[37] Even a quarter century
after the defeats at Jena and Auerstädt, the memories of these battles
remained so painful that Clausewitz's editors refused to include these
documents in his collected works, and it would be 56 years after his death
before this treatise was published. [38]

The shock of these defeats also impressed upon Clausewitz the importance of psychological and moral factors in the conduct of war. Indeed,

his theoretical approach to these forces in *On War* can be directly traced
to the calamities he witnessed in 1806:

> The scale of a victory does not increase simply at a rate commen-
> surate with the increase in size of defeated armies, but progres-
> sively. The outcome of a major battle has a greater psychological
> effect on the loser than the winner. This, in turn, gives rise to
> additional loss of material strength [through abandonment of
> weapons in a retreat or desertions from the army], which is echoed
> in loss of morale; the other two become mutually interactive as
> each enhances and intensifies the other.[39]

Clausewitz clearly understood the crushing psychological toll of wars
and built a theory to describe this primal human force.

Although it may seem hyperbolic, the importance of the 1806 defeats
in shaping Clausewitz's life and work is difficult to overstate. They
shattered the old Prussian order, forced him to think and write with a
sense of fanatical purpose, showed him new modes of warfare, inspired
him to become a military reformer, and ultimately led him to temporarily
abandon his country.

Military Reformer

In the spring of 1808, Clausewitz traveled to Königsberg to rejoin Scharnhorst as his aide-de-camp. Although his attachment to Prince August had
been socially prestigious and had given him ample time for intellectual
pursuits, he was happy to leave the self-indulgent monarch and to serve
his old mentor.[40] In his role as Scharnhorst's aide, Clausewitz now had
an insider's view of the radical transformation of the Prussian military.

Scharnhorst had been appointed by the Prussian monarchy to study
the recent military disasters. As the head of the Military Reorganization
Commission, Scharnhorst was given broad latitude to investigate the
recent Prussian defeat and propose reforms.[41] This was a paradigm
shift in the notoriously hierarchical Prussian military, and Scharnhorst
wasted no time in acting. He filled the newly formed commission with

like-minded reformers, including Clausewitz, many of whom he had mentored during their time at the war college. Clausewitz was never an official member of the commission, but as an aide to Scharnhorst, he had significant informal influence.[42] Because of his sharp mind and strong work ethic, he was seen as a valuable member of the broader reform movement and was included in their discussions on an ad hoc basis.

While the Military Reorganization Commission correctly identified deep structural flaws in the Prussian political and military system, it had to proceed cautiously to avoid alienating the proud military elite or the insecure monarch Frederick Wilhelm III. To move beyond merely adopting French drill and tactical methods or removing a handful of ineffective officers, the commission needed to win over the deeply conservative Prussian power brokers and secure their support for reform.[43] As a participant in these discussions, Clausewitz had an insider's view of the interplay between military and political factors that shape policy. In many ways, the commission had to subordinate its preferred military objectives to prevailing political realities, a compromise that would influence Clausewitz's thinking on civil-military relations.

Scharnhorst was a shrewd political strategist who employed a mixture of flattery, intelligence, and humor to win over many members of the Prussian elites. In addition, Scharnhorst emphasized the continuity of the proposed reforms to avoid upsetting proud military tradition or seeming to disregard the legacy of Frederick the Great.[44] Thanks to these efforts, the Prussian army was able to adopt many of the best features of the Napoleonic armies, such as improved tactics, the *levée en masse*, the expansion of the officer ranks, the creation of a national guard, and the elimination of draconian punishments. At the same time, it reinvigorated the martial spirit and traditions so highly prized by the traditionalists.[45] While few outside of Scharnhorst's inner circle of reformers recognized their full significance at the time, these accomplishments were nonetheless substantial. In a few short years, the Prussian army was transformed from one of the most outdated and ineffective militaries in Europe to one of the most modern and efficient. Prussian victories in 1813 and 1815 would soon demonstrate the effectiveness of the commission's reforms.[46] Although the Prussian and later German militaries would undergo numerous additional reforms and reinventions, the foundation had been laid for over a century of German tactical and operational dominance.

During this period, Clausewitz developed a friendship with General August von Gneisenau, another key leader in this Prussian military renaissance. The two had first met during Clausewitz's time at the war college, though they remained little more than professional acquaintances. Now working together on military reform—an issue they both cared deeply about—their relationship quickly developed into a lasting friendship. This relationship proved critical to Clausewitz as Gneisenau became the leading figure in the reform movement following Scharnhorst's death in 1813. Like Scharnhorst, Gneisenau supported and guided the reserved Clausewitz, helping to shape his career over the following decades.[47]

Despite his important work as an aide to Scharnhorst, Clausewitz sought to engage Napoleon in active combat. However, he was unable to do so because of the conditions of peace imposed on Prussia after its recent defeats. Although Scharnhorst and the military reformers were committed to restoring Prussian power, they recognized that resuming the conflict with France was not feasible at the time. Clausewitz was an astute strategist who recognized that this cautious approach was essential for Prussia's survival. At the same time, he believed that Napoleon was untrustworthy and emotionally volatile. Napoleon's ego and thirst for power were such that it was only a matter of time before he once again turned his armies against Prussia, regardless of peace treaties or the credibility of his promises.

In January 1810, Napoleon underscored Prussia's vulnerability by demanding immediate payment of the reparations stipulated in the 1807 Treaty of Tilsit. Friedrich Wilhelm and the Prussian diplomats pleaded for more time, but Napoleon refused to yield. He gave the insolvent

Prussians two options to avoid war: cede the territory of Silesia or raise the money by reducing the army to a royal bodyguard of only 6,000 men. After Scharnhorst objected to cutting the military and advocated ceding territory instead, Napoleon issued an additional demand: Scharnhorst must resign from the war ministry.[48] For Clausewitz, this confirmed that his greatest fear was true. Napoleon harbored malicious intentions toward Prussia and sought to derail the military reforms before they could be fully implemented.[49]

Bowing to French pressure, Scharnhorst was soon removed as head of the war ministry.[50] As Scharnhorst's aide, Clausewitz suddenly found himself with little to do. Although he would have preferred to remain near his mentor, Clausewitz was not permitted to do so. Instead, and much to his dismay, he was promoted and assigned to teach at the war college.

War College Professor

In August 1810, Clausewitz was promoted to major and assigned to teach at the war college as an instructor of tactics.[51] Although this assignment would further his career as both a soldier and a thinker, he was dismayed by the news. He wanted to remain in Scharnhorst's inner circle, rather than accept a position that would confine him to the classroom and remove him from familiar duties. His lack of enthusiasm was evident in the several months he delayed reporting for duty and in the complaints recorded in his private letters. He took some comfort in returning to Berlin, where his sweetheart, Marie von Brühl, lived. Yet, even this brought concern as he lacked the money or the social standing to gain her mother's approval for marriage.

Despite his initial hesitance, Clausewitz understood that a position as a war college professor gave him the opportunity to refine his own thinking on war while passing his ideas to a new generation of Prussian officers. He was tasked with teaching a course on small wars, and he dedicated himself to mastering the subject and adapting to his new responsibilities.[52] Two hundred and forty-four handwritten pages of his lectures from this period survive, and they reveal a man who was desperately trying to comprehend the defeats of 1806 and understand the rapidly evolving character of war.[53] In these lectures, which covered approximately 156 hours of class time, he argued for the use of independent commands, greater tactical and operational flexibility, and the importance of partisans in warfare.[54] He was particularly interested in the Peninsular War being fought in Spain.[55] The conflict combined partisan warfare with traditional set-piece battles, as the Spanish—supported by British regulars—fought a protracted campaign to wear down Napoleon's forces at the edge of his empire. Clausewitz recognized the campaign's importance as it unfolded and later built upon these insights in Book VI of *On War*. [56]

As far as is known, Clausewitz's teaching style differed from the theoretical approach he later employed in *On War*. He was less concerned with teaching theory, facts, or a fixed body of knowledge and more focused on building on his experiences to engage students in active learning. Much like today's staff ride method, Clausewitz required his students to assume the roles of historical commanders and make key military decisions based on what those leaders knew at the time. Clausewitz believed that this experiential approach was essential to cultivating practical judgment in his pupils.[57] Though circumstances might change, he maintained that the ability to think quickly and apply both theory and experience would serve his students well in future conflicts.[58]

While teaching at the war college, Clausewitz was also able to help write the tactically focused *Army Regulations*. This set of official instructions, published in 1812, revolutionized the Prussian military doctrine by introducing new methods of drill, marksmanship, tactical formations, and training. Although significantly shorter than other field manuals of the period, this document had a substantial impact. It was the first major revision of Prussian doctrine since the time of Frederick the Great, more than fifty years earlier, and incorporated many of the reformers' experiences from the Napoleonic Wars. As a result of Clausewitz's efforts, the Prussian army adopted a more practical set of military principles, making it one of the most modern in Europe.[59]

In addition to his other responsibilities, from October 1810 to March 1812, Clausewitz served as military tutor to the crown prince of Prussia, the future Frederick William IV.[60] Privately, he expressed frustration with his royal pupil, noting that he lacked seriousness as a soldier. Yet despite his lament of "crown-princely deaf ears," this appointment established Clausewitz as a trusted member of the elite and a rising figure in the Prussian military.[61]

A Long-Delayed Marriage

Clausewitz improved his social standing in December 1810 by wedding Countess Marie von Brühl. The couple had met seven years earlier, but they delayed marriage for practical reasons. Carl was of lower social rank, had no prospects for inheritance as a fourth son, and could rely only on a junior officer's pay. Marie's father had died the year before they met, leaving his prominent family in reduced financial circumstances.[62] Given Clausewitz's dim financial prospects, Marie's mother had understandable reservations and had blocked the union. Ultimately, Carl's promotion to major, the official blessing of the marriage by Friedrich Wilhelm III, and Marie's ability to attain an education and an income of her own made the wedding practicable.[63] After the king officially approved the marriage, Marie's mother finally gave her consent to their union. The couple wasted little time and were soon married on December 17, 1810, in a small ceremony at the *Marienkirche* in the center of Berlin.[64]

Although she was older than the typical age for marriage at the time, Marie was an ideal match for Clausewitz. She was extremely intelligent, sophisticated, socially adept, and fiercely devoted to her husband. Carl treated her as his intellectual equal, and they frequently discussed art, literature, politics, war, and philosophy.[65] The couple soon leveraged Marie's social standing and connections to move within the highest circles of Prussian society and, for a time, held considerable influence.[66]

Notes

1. Esdaile, *Napoleon's Wars*, 269–270; and Herberg-Rothe, *Clausewitz's Puzzle*, 15–38.
2. Howard, *Clausewitz*, 14–15; Lefebvre, *Napoleon*, 257; and Paret, *The Cognitive Challenge of War*, 5–16.
3. Lebow, *The Tragic Vision of Politics*, 200; and Strachan, *Clausewitz's On War*, 44–45.
4. Paret, *The Cognitive Challenge of War*, 6–19.
5. Napoleon believed that his army of 1806 was the finest he ever commanded; Paret, *The Cognitive Challenge of War*, 6; and van Creveld, *Command in War*, 58–102.
6. Stoker, *Clausewitz: His Life and Work*, 45.
7. Sumida, *Decoding Clausewitz*, 83.
8. Clausewitz, *On War* (Howard and Paret), IV.10, 254–255. For a discussion of this point, see Stoker, *Clausewitz: His Life and Work*, 54.
9. Jay Luvaass, "Student as Teacher: Clausewitz on Fredrick the Great and Napoleon" 150–170. See also Kuhn, *The Structure of Scientific Revolutions*.
10. Paret, *The Cognitive Challenge of War*, 23.
11. Howard, *The Invention of Peace*, 39.
12. This brief and undistinguished retreat was the only time Clausewitz commanded a large number of troops in battle. Stoker, *Clausewitz: His Life and Work*, 62–63; and Waldman, *War, Clausewitz, and the Trinity*, 4.
13. Esdaile, *Napoleon's Wars*, 272–277; Paret, *Clausewitz and the State*, 125–126; Strachan, *Clausewitz's On War*, 45; and Sumida, *Decoding Clausewitz*, 83.
14. Paret, *Clausewitz and the State*, 127–131; Parkinson, *Clausewitz: A Biography*, 76–80; and Stoker, *Clausewitz: His Life and Work*, 67–72.
15. Herberg-Rothe, *Clausewitz's Puzzle*, 22–24.
16. Quoted from "Unfinished Note, Presumably Written in 1830"; see Clausewitz, *On War* (Howard and Paret), 71. See also Paret, *The Cognitive Challenge of War*, 124.
17. Clausewitz, *On War* (Howard and Paret), IV.12, 267.
18. Lefebvre, *Napoleon*, 261; and Nester, *Napoleon and the Art of Diplomacy*, 194–195.
19. Esdaile, *Napoleon's Wars*, 276–277.
20. Parkinson, *Clausewitz: A Biography*, 80.

21. Herberg-Rothe, *Clausewitz's Puzzle*, 15–38; and Parkinson, *Clausewitz*, 80–81.
22. Parkinson, *Clausewitz: A Biography*, 81–83.
23. Bellinger, *Marie von Clausewitz*, 79.
24. Parkinson, *Clausewitz: A Biography*, 82–84; and Stoker, *Clausewitz: His Life and Work*, 70.
25. Bellinger, *Marie von Clausewitz*, 79.
26. Parkinson, *Clausewitz: A Biography*, 86 and 89.
27. Parkinson, 86–95; and Stoker, *Clausewitz: His Life and Work*, 70.
28. Heuser, *Reading Clausewitz*, 3.
29. Nester, *Napoleon and the Art of Diplomacy*, 203; and Sumida, *Decoding Clausewitz*, 83.
30. Lefebvre, *Napoleon*, 26 and 260.
31. Paret, *The Cognitive Challenge of War*, 77.
32. Lebow, *The Tragic Vision of Politics*, 201; and Levinger, *Enlightened Nationalism*, 44.
33. Levinger, *Enlightened Nationalism*, 44.
34. Howard, *Clausewitz*, 9.
35. Paret, *The Cognitive Challenge of War*, 7, 106, and 113–114.
36. Paret, 74–75.
37. Clausewitz, *Carl von Clausewitz: Historical and Political Writings* (Paret and Moran), 41.
38. Paret, *The Cognitive Challenge of War*, 73–74.
39. Clausewitz, *On War* (Howard and Paret), IV.10, 253. For an interesting application of this insight, see Jervis, *System Effects*, 35 and 164.
40. Parkinson, *Clausewitz: A Biography*, 108–111.
41. Clark, *Iron Kingdom*, 325; Echevarria, *Clausewitz and Contemporary War*, 45; Paret, *The Cognitive Challenge of War*, 84–86; and White, *The Enlightened Soldier*, 131–133.
42. Parkinson, *Clausewitz: A Biography*, 97; and Stoker, *Clausewitz: His Life and Work*, 75–79.
43. Herberg-Rothe, *Clausewitz's Puzzle*, 19–21; and Parkinson, *Clausewitz: A Biography*, 100–102.
44. The very first of Frederick the Great's 1747 *Instructions to His Generals* directly addresses the challenge posed by the social stratification of Prussia and the corresponding need for harsh discipline. He notes that approximately half of the Prussian army was foreign-born and that nearly all enlisted soldiers came from the lowest rungs of society. His solution to prevent desertion and to instill a fearless warrior spirt among

the ranks was strict discipline, a point not lost on Clausewitz and his fellow miliary reformers. See Frederick the Great, *Instructions to His Generals* (Phillips), 21–23 and 29. See also Paret, *Clausewitz and the State*, 59.
45. Paret, *The Cognitive Challenge of War*, 86–101.
46. Paret, 103.
47. Clark, *Iron Kingdom*, 325–327; Parkinson, *Clausewitz: A Biography*, 97;
and Stoker, *Clausewitz: His Life and Work*, 83.
48. Paret, *Clausewitz in His Time*, 100.
49. Parkinson, *Clausewitz: A Biography*, 121–123.
50. Paret, *Clausewitz in His Time*, 100.
51. Bellinger, *Marie von Clausewitz*, 104; and Stoker, *Clausewitz: His Life and
Work*, 85.
52. Stoker, *Clausewitz: His Life and Work*, 84–85.
53. Clausewitz, *Clausewitz On Small War* (Daase and Davis), 19–168. See
also Paret, *Clausewitz in His Time*, 101–102; and Paret, "Clausewitz: 'Half
against my will, I have become a Professor,'" 591–601.
54. Clausewitz did not address revolutionary or guerrilla warfare in any sustained way, referring to these themes only in passing during his lectures.
Although he later advocated for partisan warfare against France, there
is no evidence that he promoted such a course in his teaching or was
involved in preparing a resistance movement; Paret, *Clausewitz in His
Time*, 101–109; Paret, *Clausewitz and the State*, 145–187; Paret, "Clausewitz: 'Half against my will, I have become a Professor'"; and Parkinson,
Clausewitz: A Biography, 126–127.
55. Clausewitz was also interested in the popular uprising against Napoleon
in the Tyrol. Paret, *The Cognitive Challenge of War*, 97.
56. Clausewitz, *On War* (Howard and Paret), VI., 357–501. For a discussion of
how Clausewitz incorporated these insights into his work, see Echevarria, *Clausewitz and Contemporary War*, 45 and 137.
57. On this point, Clausewitz took particular inspiration from his mentor
Scharnhorst. Lebow, *The Tragic Vision of Politics*, 179; and White, *The
Enlightened Soldier*, 98–99.
58. Robertson, *The Staff Ride* and Paret, *Clausewitz and the State*, 200.
59. Parkinson, *Clausewitz: A Biography*, 128–129.
60. Echevarria, *Clausewitz and Contemporary War*, 45.
61. Paret, *Clausewitz and the State*, 201.
62. Clausewitz's brother-in-law, for example, was the director of the Royal
Theatre and a member of the Prussian court although he occasionally

provoked the anger of the sovereign with his choice of productions.
Paret, *The Cognitive Challenge of War*, 62.

63. Paret, *Clausewitz and the State*, 209.
64. Bellinger, *Marie von Clausewitz*, 106–107.
65. Paret, *Clausewitz and the State*, 103–104.
66. Strachan, *Clausewitz's On War*, 91.

CHAPTER 4

EXILE FOR PATRIOTISM: CLAUSEWITZ'S DEFECTION TO THE RUSSIAN ARMY

Joining the Russian Army

By early 1812, Clausewitz appeared to be thriving. He was happily married and enjoying newfound social prestige, and his path to high rank and influence seemed assured—yet despite these personal triumphs, Clausewitz despaired. Ever since his release from French captivity, he had been eager to resume his fight against Napoleon. While in captivity, he had developed a deep enmity for Napoleon and wanted nothing more than to end his reign as Europe's dominant military ruler. Perhaps forgetting that military aims should be subordinate to political calculations, he believed that Prussian honor required continued resistance against Napoleon, a belief that put him in direct conflict with his sovereign, Frederick Wilhelm III. Wilhelm also wanted to fight Napoleon but believed that Prussia was too weak to do so and therefore had to act cautiously. This was a prudent policy, but it did not satisfy the ambitious and patriotic Clausewitz.[1]

As early as 1811, Clausewitz worked with his friend and mentor Gneisenau to draft a plan for a popular uprising against the French. Both men had been inspired by the examples of the Spanish and Tyroleans. Clausewitz argued that Silesia was the ideal place to begin a revolution.[2] While both Clausewitz and Gneisenau believed that this plan would fail, they convinced themselves that Prussian honor demanded continued resistance. Clausewitz wrote to Gneisenau that he would, "happily sacrifice my own life" for the Prussian nation.[3] While Gneisenau ultimately chose not to implement these plans for insurrection, Clausewitz was sincere in his desire to fight. In *On War*, he blamed Frederick Wilhelm III for failing to incite a populist uprising and equated a nation's failure to commit to a people's war as the true sign of moral decay and total defeat.[4]

While this critique was somewhat unfair, it truly expressed Clausewitz's feelings on the matter. As a realist and a nationalist, Clausewitz believed it was both dangerous and degrading to rely on the benevolence of Napoleonic France for survival.[5] Clausewitz's sense of urgency intensified in February of 1812, when Napoleon demanded that Prussia provide 20,000 troops for service in the French army. This seemed to confirm that Prussia was at the mercy of the French emperor and raised the possibility that Clausewitz would be compelled to serve under Napoleon. This was likely the moment when Clausewitz resolved to actively oppose Napoleon.[6] Time was running out and, Clausewitz believed that he had a duty to resist Napoleon—the Treaty of Tilsit, the wishes of his sovereign, and diplomatic formalities notwithstanding.

To resist Napoleon, Clausewitz had to look beyond his native Prussia. An opportunity arose in 1812 when Napoleon declared war on Russia because of its refusal to comply with the blockade against Great Britain. The Russian military was undertaking its own reforms, but with less success than its Prussian counterpart. While vast in size, the Russian army was disorganized and lacked the bureaucratic, technological, and financial elements to make it an effective fighting force. It also lacked enough trained officers to command its troops and advise the tsar. With no time to develop these leaders internally, the tsar turned to foreign recruitment. While this arrangement may strike contemporary readers as unusual, the Russian monarchy had used foreign officers to augment its ranks for centuries, including notable figures as John Paul Jones, Patrick Gordon, and Michael Andreas Barclay de Tolly.

Amid these tensions, Clausewitz and about three hundred other Prussian officers, roughly a quarter of the officer corps, resigned their commissions.[7] Unwilling to resign and remain passive, Clausewitz was one of approximately thirty Prussian officers who accepted a commission in the Russian army.[8] This substantial wave of resignations and defections created a crisis within the Prussian government. Not only was the army's manpower greatly diminished, but the loyalty of its officer corps and the aristocracy was also cast into serious doubt. This sense of shock and betrayal reverberated throughout the Prussian elite and had profound consequences for Clausewitz's career.[9]

For both practical and patriotic reasons, Clausewitz harbored significant reservations about serving in a foreign military. He spoke almost no Russian; his native country was officially at peace with France; and joining the Russian army would require a formal break with his monarch, Frederick Wilhelm III, as well as a departure from Prussia just as his social and professional networks were beginning to yield tangible benefits.[10] Although joining the Russian military came at considerable personal cost, Clausewitz was a committed Prussian patriot who remained determined to oppose Napoleon. Even though this decision would mean temporarily abandoning his homeland, Clausewitz believed the larger interests of Prussia demanded this sacrifice. In a private letter to Marie, he wrote, "The standard which I've followed with love and devotion for twenty years, I am no longer allowed to bear...A melancholy feeling did indeed slowly come over me with these ideas, but it did not sadden me."[11]

To document his thinking and assert his patriotism, Clausewitz composed three political manifestos or "confessions" in February 1812 and mailed them to his friend and mentor Gneisenau. In these pages,

Clausewitz declared—using notably florid language—his love of country,
his hatred for France, and his belief that he was acting with the highest
moral principle.[12] In one especially ornate passage, he described his
devotion to his country and the imperative to act boldly:

> I believe and confess that a people can value nothing more highly
> than the dignity and liberty of its existence. That it must defend
> these to the last drop of blood. That there is no higher duty to
> fulfil, no higher law to obey. That the shameful blot of cowardly
> submission can never be erased. That this drop of poison in the
> blood of the nation is passed on to posterity, crippling and eroding
> the strength of future generations. That the honor of the king and
> government are one with the honor of the people, and the sole
> safeguard of its well-being. That a people courageously struggling
> for liberty is invincible. That even the destruction of liberty after
> a bloody and honorable struggle assures the people's rebirth. It
> is the seed of life, which one day will bring forth a new securely
> rooted tree.[13]

While these lines may strike the modern reader as a blend of revolutionary fervor and proto-nationalist rhetoric, it is clear that Clausewitz
believed he had to fully commit himself to serving the best interests of
the Prussian people.[14] In the name of liberty, he joined the Russian army
and became an exile from his native Prussia.[15] Clausewitz's application
for a commission in the Russian army was accepted in March 1812, and
because he had not received royal permission to join a foreign military,
he was officially declared an enemy of Prussia. He initially entered the
Russian army as a lieutenant colonel and was assigned to the staff of
General Karl Ludwig von Phull, a fellow Prussian expatriate who had
entered Russian service in 1807.[16]

As he prepared to depart, Clausewitz composed a short monograph
titled *Principles of War*, later referred to by historians as *Instructions for
the Crown Prince*, summarizing the strategic and tactical lessons he had
drawn from his study of war. He gave the volume as a parting gift and
final lesson to his pupil, the sixteen-year-old crown prince Frederick

William.[17] With a mixture of flattery and self-promotion, he noted that it contained, "the most important principles of the art of war to complete my course of instruction for his Royal Highness the Crown Prince."[18] Although the work was hastily written and primarily tactical in focus, it demonstrates that Clausewitz had already begun his deep intellectual study of war.[19] For example, the concept of friction was already well developed in this early work and would later be a key element of *On War*.[20]

Clausewitz left Berlin in a horse-drawn coach on March 31, 1812, the very day that French soldiers entered the city on their way toward the Russian frontier. While Clausewitz believed in the righteousness of his cause, he was overcome by physical distress. He reflected on his situation in a letter to Marie, noting that he "cursed every stone over which I drove" eastward toward Russia.[21]

Clausewitz was now an outlaw. He had not received formal permission to join Russian military and had entered the service of an enemy camp opposed to his native Prussia. Shortly after his departure, a Prussian court formerly charged Clausewitz, along with approximately thirty other officers, with treason. The court ordered him to appear to defend himself and warned him that he would be convicted in absentia if he failed to return to testify. If convicted, Clausewitz could be stripped of all his property and inheritance and would be a permanent exile from his homeland. Although Marie urged him not to be concerned with these legal matters, this development underscored the steep price that Clausewitz was paying for his resistance to Napoleon.[22]

The Russian Campaign of 1812

Now that he had cast his lot with the tsar, Clausewitz did his best to focus on the immediate tasks—improving the beleaguered Russian army and vanquishing Napoleon. Despite his inability to speak Russian and the considerable mistrust that others felt due to his Prussian origins, Clausewitz gained privileged insight into the 1812 campaign. He served in a series of staff positions and witnessed firsthand the battle of Borodino, the burning of Moscow, and Napoleon's disastrous winter retreat.[23] He sought retribution for the Prussian humiliations of 1806, and this desire drove him to take extraordinary risks. For example, in the aftermath of Borodino, he helped cover the Russian retreat and demonstrated his bravery by repeatedly exposing himself to enemy fire, even having a horse shot out from under him.[24] This was highly uncharacteristic conduct for a staff officer, yet Clausewitz fought with exceptional determination.[25]

Despite Clausewitz's commitment to the destruction of Napoleon, he was dismayed to discover that during his absence from Prussia, he had been formally charged by the Prussian state for bearing arms against his homeland. He wrote to Marie, expressing pique and resentment:

> Let them do it in God's name! Anyone who has witnessed the scenes of misery and need here, which the German governments helped bring about, will not feel his pride broken by their condemnations.[26]

Clausewitz's service in the Russian army would have profound implications for his career, but for the time being he tried to ignore these legal proceedings and concentrate on the immediate objective—the destruction of Napoleon. He sought refuge in his duties as an officer, but the physical and psychological toll proved severe. As early as August 1812, stress and exertion had begun to take a noticeable toll on his health. At the age of only thirty-two, he began to suffer from gout, toothaches, and exhaustion, while his hair began to fall out in patches and his skin became leathery.[27] Although Clausewitz was never imposing physically, this decline was extreme. For the remainder of his life, he suffered from poor health, and his frail body seemed to exacerbate his shy and aloof personality.[28]

Service in the Russian army also had a profound impact on Clausewitz's psychology. While he welcomed the French defeat, the surrounding chaos and carnage left a deeper impression on him. No longer was he the naïve youth who marched off to Jena in 1806, writing letters about the likely victory and the glories of war. Now, he was more mature and realistic in his thinking. His moods grew darker and more introspective. As he witnessed the burning of Moscow, he described it as an "act of barbarity, a result of the enemy's hate, insolence, and cruelty." He took little comfort from the subsequent French retreat from Russia and wrote Marie that such "ghastly scenes" nearly drove him "mad." He added that he could endure them only because he was a battle-hardened veteran. He later wrote, "I felt as if I could never be released from the terrible impressions of the spectacle. I only saw a small fraction of the famous retreat, but in this fraction of some three days' march, all the horrors of the movement were accumulated."[29]

Despite the strains inflicted by the Russian campaign, Clausewitz remained deeply engaged in analyzing its military lessons.[30] He quickly recognized that the vast Russian territory was key to their success. It allowed them to conduct a fighting retreat and wear down the combat power of the advancing French forces. Napoleon had calculated that he could achieve rapid victory through decisive battles, just as he had at Jena and Auerstädt.[31] By refusing to accept battle on the frontiers, the Russians invalidated Napoleon's plan for a quick campaign. When faced with this reality, Napoleon continued to drive his army past the point at which victory could be achieved. As Clausewitz described, "the invader was destroyed by his own exertions."[32]

These observations came to shape Clausewitz's views that defense is the stronger form of warfare and helped to inform his concept of the culminating point of victory.[33] He later summarized his thoughts in *On War*, noting:

> all Europe was opposed to Bonaparte; he had stretched his resources to the very limit; in Spain he was fighting a war of attrition; and the vast expanse of Russia meant that an invader's strength could be worn down to the bone in the course of five hundred miles' retreat. Tremendous things were possible; not only was a massive counterstroke a certainty if the French offensive failed (and how could it succeed if the Tsar would not make peace nor his subjects rise against him?), but the counterstroke could

bring the French to utter ruin. The highest wisdom could never
have devised a better strategy than the one the Russians followed
unintentionally.[34]

During this brutal campaign, Clausewitz also observed the deep commitment and mobilizing spirit of the Russian people and their leaders. This
passion helped motivate the troops to endure enormous sacrifices and
contributed to Russia's eventual victory. These observations helped shape
his theories on the energies of the people, the emotional dimensions of
war, and the difficulty of defeating a nation at arms.

As Napoleon began his long retreat from Moscow, Clausewitz again
observed the need to pursue and destroy a defeated force. He believed
Napoleon's forces were so thoroughly defeated that they were incapable
of fighting another battle. Final victory seemed within reach, provided
the Russians could exploit their gains and prevent Napoleon's remaining
force from escaping. In theory, the opportunity to end the Napoleonic
Wars had arisen in 1812.[35] In practice, however, the Russian army was
hindered by exhaustion, logistical challenges, and a corps of generals
whose internal disputes made coordination difficult.

The opportunity to destroy the remainder of the French army soon
passed, and Napoleon was able to reconstitute his force, drawing upon his
charisma and the passions of the French people. The fighting continued
for another three years, much to Clausewitz's disappointment. Yet this
missed opportunity helped him to better understand the gap between
theory and reality as well as the enormous complexity of military
campaigns.[36] While many of Clausewitz's views on warfare were still
evolving, this period of service laid the groundwork for some of his most
important theoretical insights.

Political Intrigue

As Napoleon retreated from Russian territory, Clausewitz began to
develop a bold plan. He believed that the time had come for his native
Prussia to abandon its alliance with France and join the war against

Napoleon. For Clausewitz, this was more than a theoretical exercise. He feared his homeland would become a site of devastation and that Napoleon would exploit it for manpower and resources for the coming campaigns. More personally, Clausewitz had also learned that his brother Frederick was serving in a Prussian unit directly in the path of the advancing Russian forces and was determined to prevent a possible fratricidal confrontation.[37]

To avert these potential disasters, Clausewitz again placed himself in a position to undermine Prussia's official policy. He began by sending messages to the Prussian General Ludwig Yorck, urging him to abandon the alliance with France, change allegiance, and defend the true interests of his homeland. Although Yorck had been in secret contact with the Russians since September, he understood the need to act with extreme discretion.[38] Yorck also wanted to avoid fighting the advancing Russians, and had even considered making a similar offer, but he was waiting to receive orders from the king. He hoped the Prussian monarch would seize this opportunity to strike a decisive blow against Napoleon, but he was disappointed by the unwillingness of the king and his advisors to alter their position. Since his orders were to resist the Russian forces, Yorck refused to acknowledge Clausewitz's overtures as a means of maintaining plausible deniability.[39]

After his covert messages went unanswered, Clausewitz ordered one of his fellow officers to ride ahead and arrange a temporary cease-fire between the advancing Russians and the Prussian force under the command of General Yorck. On December 29, a truce was quickly arranged, and Clausewitz immediately placed himself at the center of the negotiations. He served as an interlocutor between three men he knew from his days as a Prussian officer, the Prussians under Yorck and Friedrich Emil Ferdinand von Kleist, and the Russian forces under the Prussian-born Hans Karl von Diebitsch.[40] Tensions were high, and the stakes considerable. Yorck and Kleist commanded over 14,000 troops that were deployed to the Prussian border with orders to resist any invasion.

This force could have blocked the Russian pursuit of Napoleon, allowed him to escape, and escalated the war by instigating a new round of fighting between Prussia and Russia.[41] To avoid this, the two Prussian expatriates, Clausewitz and Diebitsch, were essentially asking Yorck and Kleist to join them in committing treason by defecting to serve the broader interest of their nation.

As the fate of Prussia and the course of the Napoleonic Wars hung in the balance, Clausewitz proved indispensable in these negotiations. He was not only politically savvy but also personally familiar to General Yorck. As a young officer serving as Scharnhorst's aide, Clausewitz had frequently worked with Yorck, and the two had recently collaborated on the 1812 revision of the army regulations on light infantry. Privately, Clausewitz believed Yorck to be "morose, melancholic, and secretive" and noted that "Personal attachment is rather foreign to him." He also observed that "what he does, he does for the sake of his reputation and because he is naturally competent."[42]

In addition to these insights, Clausewitz held a significant advantage in the upcoming negotiations. His forces were militarily superior, and Yorck was eager to oppose Napoleon. Clausewitz was able to press Yorck into a decision by declaring that the Russians would be willing to attack with overwhelming numbers in two days' time.[43] The precise details of their negotiations have been lost to history, but the professional reputation and personal trust that they had previously established appear to have been critical in coaxing Yorck to risk his life and reputation by switching sides.[44] While pondering his options, Yorck apparently asked Clausewitz if it was true that the Russians could attack on December 31. With an honesty that acknowledged friction and uncertainty in war, Clausewitz replied, "I guarantee the honesty...whether these intentions really will be fulfilled I can, of course, not guarantee, since Your Excellency knows that in war with the best will in the world one must often fall short of the line one has set oneself." This candid reply was enough for Yorck who reportedly took Clausewitz's hand and replied, "You have me."[45]

The resulting Convention of Tauroggen, which was drafted by Clause-witz and signed by Yorck on December 30, 1812, was a political master-stroke. It stopped short of an official alliance against France, but it ended the fighting between Prussia and Russia for two months, forced Napoleon to continue his retreat westward rather than staging a defensive stand in Prussian territory, and allowed Russian forces to pass uncontested.[46] Clausewitz permitted himself a note of pride in describing this accom-plishment in a letter to Gneisenau, "We have, however, made a campaign that certainly ranks among the most original in the history of warfare, and which perhaps will be the most momentous in the annals of nations."[47]

Despite their success, the king of Prussia had not approved the agree-ment and viewed it as a treasonous subversion of his authority. The monarch's authority was further challenged when the East Prussian Estates met in a special session and voted to reinforce Yorck's forces with 20,000 *Landsturm* militia.[48] Clausewitz quickly began planning to deploy these forces to attack Napoleon, but ultimately these troops never materialized and the operation was abandoned.[49] While the king was privately happy to have avoided unnecessary bloodshed, publicly, he needed to reassert his own sovereignty and to avoid a war with Napoleon. He ordered the arrest and court-martial of Yorck and publicly denounced the Convention of Tauroggen and the declarations of the East Prussian Estates.[50] For now, the king was unready to challenge Napoleon, mistrusted Russia, and was fearful that he could not control the populist uprisings that a war with Napoleon would incite.[51]

Personally, the king believed that his authority had been undermined by his own generals and that Clausewitz had been the principal instigator. For Wilhelm, this political intrigue marked Clausewitz as a disloyal conniver, a reputation that he never fully overcame. This mistrust would have profound long-term consequences for Clausewitz's career, but for the moment, he savored the private triumph. Like most Prussian nationalists, Clausewitz viewed this treaty as the first step toward restoring national dignity by nullifying the Treaty of Tilsit and avoiding a fratricidal campaign against his brother and his fellow countrymen. In his typically understated manner, Clausewitz claimed that Tauroggen, "in all likelihood very considerably speeded up the final outcome" of the defeat of Napoleon.[52]

After Tauroggen, Clausewitz returned to Berlin expecting to be welcomed as a national hero and given a prominent command in the Prussian army. That expectation, however, was not fulfilled. In fact, on March 11, 1813, the very day Wittgenstein's forces entered Berlin, a court met to convict him on the charges of joining the Russian army without permission. The king, who had refused to defend Berlin and fled the city, wanted to make an example out of him for his perceived disloyalty.[53] Lacking better options, Clausewitz did not appear in court, but instead pleaded his case with the Empress Marianne who had remained in Berlin to care for a sick child. The empress had become something of a popular hero for her loyalty to the city but was unable to help. Even though the empress was enthralled by Clausewitz's exploits, she had little influence over such matters.[54] Clausewitz next had his mentor Scharnhorst appeal to the king, but despite this powerful ally, his application to rejoin Prussian service was officially rejected. He was given a vague promise that his application would be reconsidered if he could achieve some great distinction fighting against France, but this did little to comfort the dejected Clausewitz.[55]

Advisor to General Blücher

With few other options, Clausewitz joined the Russian-German Legion in early 1813. This unit was part of the Russian tsar's army but was composed primarily of Prussian expatriates who had joined to fight Napoleon. Clausewitz was assigned as a liaison officer to Marshal Gebhard von Blücher of the Prussian army and fought with his forces in the Battle of Großgörschen (also known as Battle of Lützen) in May 1813.[56]

Desperate to prove his worth, Clausewitz was a relentless presence during the Battle of Großgörschen. Starting at approximately 2:00 a.m.

on May 2, he mounted his horse and began one of the most difficult and dangerous days of his life. He carried messages, endured a devastating artillery barrage, and even personally led troops into battle. While it was not the typical duty for a liaison officer, he was tasked with leading a detachment of Prussian cavalry, a mission that he eagerly accepted. Clausewitz and his men were soon surrounded by French infantrymen and a fierce and confusing melee ensued. As the French foot soldiers began to pull the Prussians from their horses and bayonet them as they lay helpless on the ground, Clausewitz took charge. Fighting back with his sword, he led his men out from the trap, cutting a path through the attackers. He received a minor bayonet wound behind his right ear but otherwise emerged unscathed.[57] Clausewitz quickly reformed his men and led a counterattack back into the battle he had just escaped. Although the French ultimately won the Battle of Großgörschen, Clausewitz's brave actions helped stabilize the Prussian position and avert total collapse.[58]

As was so often the case in Clausewitz's life, this triumph was met with a corresponding tragedy. In June 1813, he learned that his mentor, Scharnhorst, had died as a result of wounds suffered at Großgörschen.[59] This was especially painful, as the injury—a gunshot wound to the foot —had initially seemed minor, and a full recovery had been expected.[60] In Clausewitz's mind, Scharnhorst neglected his recovery in order to continue working, and Clausewitz saw his death as a testament to his unrelenting dedication and patriotism.[61]

Without his father figure and patron, Clausewitz despaired.[62] He feared that, without Scharnhorst's support, he would flounder and that his enemies would deny him the return to Prussian service he so passionately desired. He believed that his heroics at Großgörschen had amply demonstrated his worth, and he again petitioned the king for readmission into the Prussian service. He had some reason for optimism as his friend and mentor General Gneisenau had replaced Scharnhorst as chief of staff and requested Clausewitz by name for a position on his staff.[63] However, despite this direct appeal, Clausewitz's hopes were soon crushed. The king rejected this request, and Clausewitz remained in Russian service.[64]

After the Battle of Großgörschen, the German Legion was attached to the Army of the North and stationed in a quiet, northern sector of Prussia. Because of its geographic isolation, the Army of the North saw little action and Clausewitz again worried that he would never have the opportunity to earn the redemption.[65] His frustrations were exacerbated by his poor health, which having never recovered from the rigors of the 1812 campaign, continued to deteriorate.[66] He now found himself isolated—without a mentor, a country, meaningful action, or his health.

Back to Prussian Service

On April 14, 1814, Clausewitz was allowed to re-enter Prussian service but even this return came with ambivalence. By this time, Napoleon had been expelled from Germany and exiled to Elba, and there appeared to be little prospect for renewed conflict. Clausewitz was gratified that Napoleon had been defeated and was proud to again wear the Prussian uniform, yet he regretted playing only a limited role in this victory. For both personal and strategic reasons, Clausewitz supported Blücher and other hawks who advocated for a continuance of the war after Napoleon's exile in 1814. Clausewitz sought the opportunity to fight for Prussia, but he also recognized that France possessed the economic and demographic capacity to rapidly reconstitute its forces. He believed it was essential to eliminate France's future military potential, but there appeared to be little appetite for continued military action.

Political considerations soon superseded military ones as cracks began to form in the Allied coalition. War weariness among the Austrian leadership led to demands for a rapid end to hostilities following Napoleon's abdication. These demands fractured the coalition and rendered a unified policy against France increasingly untenable. Drawing on these developments, Clausewitz would later emphasize the importance of achieving national war aims and the political complexities of coalition warfare in his writings. For now, however, his warnings were unheeded. Not only was Clausewitz a relatively junior officer who had lost his mentor, but he was also viewed by some as untrustworthy and unreliable.[67]

Napoleon's Return and a Chance for Revenge

Napoleon's escape from Elba and his climactic return to power was a shock to Europe. In retrospect, Napoleon's latest bid for power was an act of desperation with little chance for success, but in the moment a palpable sense of crisis gripped Europe.[68] The Congress of Vienna declared Napoleon an outlaw and rapidly reassembled a coalition to confront their old nemesis. Although many European leaders responded with alarm, Clausewitz privately appreciated Napoleon's return. The protean French general not only proved Clausewitz's dire warnings regarding the termination of the war in 1814 correct, but it also gave him the opportunity to confront his adversary once again—this time as a Prussian officer.

In the months that followed, Clausewitz served with distinction as the chief of staff for General Johann von Thielmann, commander of the Prussian III Corps. In this capacity, Clausewitz contributed significantly to the battles of Ligny and Wavre during the Hundred Days.[69] While not as famous as Waterloo, the battles of Ligny and Wavre were essential to Napoleon's final defeat. At Ligny, coalition forces attempted to trap Napoleon's Army of the North between converging allied armies. As he had done at Jena and Auerstädt, Napoleon responded with characteristic speed and used his interior lines to launch coordinated attacks. On June 16, 1815, he struck simultaneously on two fronts, inflicting severe losses on Blücher's Prussians at Ligny and nearly defeating Wellington's forces at Quatre Bras. Although the French had won tactical victories in both battles, they failed to destroy either allied army and suffered losses they could ill afford.

After failing to achieve a decisive victory at either Ligny or Quatre Bras, Napoleon was forced to attack Wellington's force at Waterloo on June 18.

He knew that the Prussian army remained a threat to his rear and hoped to defeat the British before his enemies could unite and bring their superior numbers to bear. For Clausewitz, Napoleon's actions were those of a "desperate gambler, indifferent to all rational calculations," yet the French once again came dangerously close to prevailing.[70] Fortunately for the Allies, General Blücher arrived with reinforcements and counterattacked the exhausted French forces, driving them from the field and securing the coalition's ultimate victory.

At Wavre, on June 18–19, the Prussians again contributed to the success of the campaign by serving as a blocking force that kept a detachment of approximately 30,000 French reinforcements from reaching Napoleon at Waterloo. Despite being outnumbered, the Prussian III Corps held for two days and bought time for Wellington and Blücher. As a result of the III Corps' stand at Wavre, Blücher was able to release some 50,000 Prussian troops to fight at Waterloo and play a decisive role in the final outcome. While the tactical results of Wavre were inconclusive, the strategic impact was immeasurable as it allowed the allies to concentrate their forces while preventing Napoleon from doing the same.

Eager for redemption, Clausewitz fought with a fierce passion at both Ligny and Wavre. He won the respect of his peers and was indispensable to the Prussian victory, yet he believed that his courageous efforts had gone largely unnoticed by Blücher and the Prussian elite. He remained convinced throughout his life that the heroic efforts by the Prussians at Ligny and Wavre had been minimized by a combination of British historians and General Blücher.[71] The British naturally emphasized their countrymen's actions at Waterloo while Blücher highlighted his role as a redeemer after he recovered from a serious injury at Ligny and arrived at Waterloo just in time for the critical counterattack that decisively defeated Napoleon. Despite this personal bitterness, Clausewitz had reason to be extremely proud of his efforts on the fields of Ligny and Wavre. He had served bravely and played a critical role in the defeat of Napoleon.[72] Shortly after the campaign, Clausewitz's commanding officer, General von Thielmann, noted, "He fills his post with distinction...a man of just as much intellectual as moral value and on the general staff [he] is fully in his element."[73]

Despite these personal accomplishments, Clausewitz was deeply discouraged. In the immediate aftermath of the fighting at Waterloo, Clausewitz lamented that the French army was allowed to retreat in relatively good order toward Paris. He argued for a more vigorous pursuit of the retreating foe, but his calls for continued pressure were largely ignored by the Allied leadership. Clausewitz understood that Napoleon's charisma, when combined with the enduring appeal of nationalism, remained a potent force, and he believed that only the capture of Paris and the imprisonment of Napoleon would end the war.[74] These fears appeared to be coming true as Napoleon abandoned his army and fled to Paris.[75] Despite his attempts to rekindle patriotic fervor, Napoleon's desperate attempt to raise another army failed and he soon abdicated in favor of his four-year-old son, Napoleon II. After a few skirmishes on the outskirts of the city, the allied forces captured Paris on July 7 and Louis XVIII was restored to the monarchy the next day. Napoleon became a fugitive but surrendered to the British on July 15 and was exiled to the island of St. Helena where he lived until his death in 1821.

Ultimately, for Clausewitz, the end of the Napoleonic Wars was anticlimactic. He argued for more punitive surrender terms for France than many of his contemporaries, but he also understood the difficulties of maintaining the allied coalition and the limited appetite for continued conflict.[76] The Treaty of Paris and the Congress of Vienna were political compromises that, though imperfect, helped establish a durable and lasting peace for Europe by balancing the interests of all parties. While Clausewitz could not know it at the time, he had fought his last battle.

NOTES

1. Bellinger, *Marie von Clausewitz*, 112.
2. Paret, *Clausewitz and the State*, 214; Stoker, *Clausewitz: His Life and Work*, 87; and Strachan, *Clausewitz's On War*, 182.
3. Strachan, *Clausewitz's On War*, 52.
4. Clausewitz, *On War* (Howard and Paret), VI.26, 483. See also Sumida, *Decoding Clausewitz*, 165.
5. Paret, *Clausewitz and the State*, 214.
6. Bellinger, *Marie von Clausewitz*, 115–117; and Parkinson, *Clausewitz: A Biography*, 133–134.
7. Craig, *The Politics of the Prussian Army*, 58.
8. Paret, *Clausewitz and the State*, 220.
9. Clausewitz's defection from his native Prussia would, in a modern civil–military context, be regarded as nothing short of treason. Although he believed he was acting in the best interests of both his king and his country, his actions stand in stark contrast to the apolitical servant-leader model now expected of professional military officers; Huntington, *The Soldier and the State*, 1957.
10. Clausewitz hired a personal servant, Jascha, who was of Russian-Jewish descent and served as his translator and personal valet for the rest of Clausewitz's life. Bellinger, *Marie von Clausewitz*, 126.
11. Stoker, *Clausewitz: His Life and Work* 102.
12. Clausewitz, *Clausewitz On Small War*, 169–216.
13. Strachan, *Clausewitz's On War*, 53–54 and 128. See also Paret, *Clausewitz and the State*, 216–218. For a slightly different translation of this passage, see Clausewitz, *Clausewitz On Small War*, 173.
14. It has been argued that this represents Clausewitz's earliest articulation of the primacy of the political; Hagemann, *Revisiting Prussia's Wars Against Napoleon*, 133.
15. For a discussion of the experience of exile in the writings of Thucydides, Machiavelli, and Clausewitz, see Aron, *Clausewitz: Philosopher of War*, 12; and Lebow, *The Tragic Vision of Politics*, 171 and 188.
16. For his part, Clausewitz had a rather low opinion of General Phull, believing him honest but inefficient; Paret, *Clausewitz and the State*, 222–223.

17. During World War II, this document was published as a quick primer on Clausewitz's thinking. Unfortunately, this led to significant misunderstandings as the work was written quickly for a juvenile audience and does not reflect the mature arguments of *On War*; Paret, *Clausewitz in His Time*, 34–35.
18. Bassford, *Clausewitz in English*, 10.
19. Interestingly, the tactical focus of *Instructions* may have resulted in it having a greater influence on pre–World War I British military thought than the longer and broader work, *On War*. Bassford, *Clausewitz in English*, 107; and Stoker, *Clausewitz: His Life and Work*, 99.
20. Bassford, *Clausewitz in English*, 10.
21. Bellinger, *Marie von Clausewitz*, 117.
22. Bellinger, 117 and 123–124.
23. Paret, *Clausewitz and the State*, 225.
24. Stoker, *Clausewitz: His Life and Work*, 25.
25. Paret, *Clausewitz and the State*, 225.
26. Paret, 226.
27. Stoker, *Clausewitz: His Life and Work*, 117; and Strachan, *Clausewitz's On War*, 56.
28. Clausewitz's reserved nature, combined with his ruddy complexion, led some detractors to suggest he was an alcoholic, claiming that the redness in his cheeks came from starting each day with a bottle of wine. A more likely explanation is that his permanently flushed appearance was the result of frostbite sustained during the Russian campaign of 1812, rather than the effects of excessive drinking; Heuser, *Reading Clausewitz*, 4; and Stoker, *Clausewitz: His Life and Work*, 138.
29. Parkinson, *Clausewitz: A Biography*, 175 and 194.
30. Herberg-Rothe, *Clausewitz's Puzzle*, 2 and 15–38; and Paret, *Clausewitz and the State*, 222.
31. Lefebvre, *Napoleon*, 226; and Stoker, *Clausewitz: His Life and Work*, 105–106.
32. Clausewitz, *On War* (Howard and Paret), VI.8, 385. Clausewitz was also deeply impressed by Napoleon's skillful use of defensive operations during the 1814 campaign. See Herberg-Rothe, *Clausewitz's Puzzle*, 28; and Stoker, *Clausewitz: His Life and Work*, 196.
33. These views appear to have been further shaped by Napoleon's defensive campaigns of 1814, which—though brilliant—ultimately yielded him less than he might have expected through negotiation; Stoker, *Clausewitz: His Life and Work*, 192.

34. Clausewitz, *On War* (Howard and Paret),VIII.8, 615. See also Paret, *Clausewitz and the State*, 224.
35. Sumida, *Decoding Clausewitz*, 90.
36. Parkinson, *Clausewitz: A Biography*, 174.
37. Parkinson, 201.
38. Paret, *Clausewitz and the State*, 229.
39. Paret, *Yorck and the Era of Prussian Reform, 1807–1815*, 192.
40. Paret, *Clausewitz and the State*, 230.
41. Paret, *Yorck and the Era of Prussian Reform*, 192.
42. Stoker, *Clausewitz: His Life and Work*, 139.
43. Paret, *Clausewitz and the State*, 230.
44. Aron, *Clausewitz: Philosopher of War*, 29; Paret, *Yorck and the Era of Prussian Reform*, 125–126, 171, and 179; and Parkinson, *Clausewitz: A Biography*, 201–203.
45. The passion of this scene reminded Clausewitz of the philosopher Friedrich Schiller's *Wallenstein*. Paret, *Clausewitz and the State*, 230; and Stoker, *Clausewitz*, 141.
46. Clark, *Iron Kingdom*, 358–359; and Paret, *The Cognitive Challenge of War*, 54–55.
47. Stoker, *Clausewitz: His Life and Work*, 143.
48. Paret, *Clausewitz and the State*, 231.
49. Bellinger, *Marie von Clausewitz*, 130.
50. Esdaile, *Napoleon's Wars*, 498; and Stoker, *Clausewitz: His Life and Work*, 145.
51. Esdaile, *Napoleon's Wars*, 498.
52. Paret, *Yorck and the Era of Prussian Reform*, 195.
53. Bellinger, *Marie von Clausewitz*, 132.
54. Bellinger, 32–133.
55. Paret, *Clausewitz and the State*, 232; and Parkinson, *Clausewitz: A Biography*, 213.
56. Paret, *Clausewitz and the State*, 232; and Echevarria, *Clausewitz and Contemporary War*, 45. Interestingly, Clausewitz was replacing his friend and former war college classmate Karl von Tiedemann, who had been killed on August 22; Stoker, *Clausewitz: His Life and Work*, 132.
57. Stoker, *Clausewitz: His Life and Work*, 158.
58. Parkinson, *Clausewitz: A Biography*, 218–220; and Strachan, *Clausewitz's On War*, 61.
59. Paret, *The Cognitive Challenge of War*, 103.

60. Stoker, *Clausewitz: His Life and Work*, 157; and White, *The Enlightened Soldier*, 162.
61. Paret, *Clausewitz and the State*, 239. Marie was also deeply saddened Scharnhorst's death, comparing it to the loss of her father more than a decade earlier; Bellinger, *Marie von Clausewitz*, 145.
62. Stoker, *Clausewitz: His Life and Work*, 165.
63. Craig, *The Politics of the Prussian Army*, 63.
64. Paret, *Clausewitz and the State*, 239.
65. Echevarria II, *Clausewitz and Contemporary War*, 46; Parkinson, *Clausewitz: A Biography*, 230–231; and Stoker, *Clausewitz: His Life and Work*, 179–180.
66. Paret, *Clausewitz and the State*, 246.
67. Howard, *Clausewitz*, 5.
68. Clark, *Iron Kingdom*, 371–372.
69. Bellinger, "A Timid Staff Officer?"
70. Clausewitz, *On Wellington* (Hofschörer), 133. See also Herberg-Rothe, *Clausewitz's Puzzle*, 36.
71. Echevarria, *Clausewitz and Contemporary War*, 46; Parkinson, *Clausewitz: A Biography*, 263–289; and Stoker, *Clausewitz: His Life and Work*, 235.
72. Paret, *Clausewitz and the State*, 248–250.
73. Stoker, *Clausewitz: His Life and Work*, 254.
74. Parkinson, *Clausewitz: A Biography*, 287.
75. Clausewitz would later write that Napoleon made his defeat total by his actions at Waterloo; yet in the moment, he feared that the protean Corsican might once again reconstitute his forces. Herberg-Rothe, *Clausewitz's Puzzle*, 35–36.
76. Parkinson, *Clausewitz: A Biography*, 286–287.

CHAPTER 5

FROM REFORM TO REFLECTION: CLAUSEWITZ'S FRUSTRATION AND THE WRITING OF ON WAR

Return to German Military Reforms

The months following the final defeat of Napoleon were among the happiest of Clausewitz's life. He was reunited with Marie, welcomed back into Prussian society, and assigned to serve directly under his friend, General Gneisenau.[1] Everything seemed to be going in his favor, and he returned to Berlin with a sense of vindication. Clausewitz was particularly eager to finish the job of reforming the Prussian military that he had abandoned three years before to enter Russian service.[2] He believed that the Napoleonic Wars had proven his theories and underscored the need to institutionalize and extend this hard-won experience.

During this period of intellectual energy and optimism, Clausewitz first began to consider writing a book. At this stage, his thoughts were rudimentary and disorganized, but he appreciated that he had a unique combination of intellectual gifts and an insider's perspective on the recent campaigns. He had had the opportunity to serve on the staffs of multiple generals and observe how they planned campaigns, made decisions, and responded to the uncertainty and stress of battle.[3] Clausewitz recognized that his access to the inner workings of the generals' minds was extremely valuable for testing best practices and theory building, and, in 1816, he began to write notes and informal studies to capture these lessons before they were lost. At this stage, he seems to have wanted to write only for military specialists and senior leaders, and it would be several more years before he embarked on the more ambitious task of writing a theory of warfare for a mass audience. Despite their limited scope, these early intellectual efforts should not be discounted. In the years immediately following the Napoleonic Wars, Clausewitz naively believed that he could reform the Prussian military, and he saw writing shorter, policy-focused pieces as a means of advancing his ideas.[4] As his wife Marie would later claim, by 1816, "the fruit that had ripened in the course of his rich experiences during four significant years of warfare," was ready for harvest.[5]

Clausewitz anticipated that his proposed reforms would be met with enthusiasm by the Prussian state. He wanted to complete the work of Scharnhorst and the military reformers and argued for a wide range of changes, from tactical reforms to a modernization of the reserves system, to better pay and a more robust meritocracy within the officer corps. The frustrating reality was that the Prussian bureaucracy saw little reason to change. The Prussian state had long been conservative and resistant to change, and now it was exhausted and nearly bankrupt. Unlike the period following the defeats at Jena and Auerstädt, there was no shock to the system or imminent threat to the state. Prussia was victorious, and there was little desire among the Prussian elites to study lessons learned or implement meaningful reforms. The hard-won fruits of victory were being allowed to rot, and future generations of Prussians would be forced to pay the price for their government's indifference.

For a fierce patriot like Clausewitz, this neglect was infuriating. Clausewitz had a gift that he was unable to fully use. What made this even crueler was that as a member of Prussian society, he had an insider's view of the inner workings of power. He understood the problem and had solutions, but he could do little about it. He angered many with his repeated demands for reform, and his increasingly sour mood marked him as something between a patriotic gadfly and an opinionated malcontent.

Now, when he should have been at his happiest, he fell into a deep depression and feeling of hopelessness. Clausewitz's life appeared to be at a dead end. He had climbed rapidly in rank and society but had no clear path for further advancement. He had distinguished himself on dozens of battlefields but would never again lead men in combat. He had hundreds of ideas about military theory and practice, but there was little interest in his work. Much to his frustration, Clausewitz's life as a man of action had ended. His life as a thinker was about to begin.

Director of the War College

Clausewitz was promoted to brigadier general and appointed director of the war college in May 1818, with strong backing from his mentor, General Gneisenau. This appeared to be an ideal match for Clausewitz's intellectual gifts, and he had enjoyed his earlier tenure as an instructor there prior to departing for Russia in 1812. In reality, however, the position proved disappointing. Clausewitz regarded it as a dead-end role with limited prospects for either promotion or field command. Behind the scenes, the king continued to question Clausewitz's loyalty and privately asked Gneisenau if he was trustworthy enough to mentor the future leaders of the Prussian army. Gneisenau expressed unwavering support, and the appointment was allowed to proceed. After a delay of some six months, Clausewitz formally assumed his post at the war college.[6]

Upon arriving at the war college, Clausewitz was dismayed to learn that his new role would be purely administrative, with no opportunity to teach. This was very disappointing for Clausewitz, who had wanted to inspire and mentor a new generation of Prussian officers, just as Scharnhorst had done for him.[7] One wonders why as the director of the school Clausewitz could not change the rules and teach at least in a limited capacity. However, this was impossible, a grim testament to the staggering rigidity and close-minded nature of the Prussian military system.

Clausewitz also found his administrative role deeply unsatisfying. He saw problems everywhere, ranging from an outdated curriculum and inadequate academic rigor to a dismissive attitude toward the practical application of military knowledge. At the time, approximately sixty percent of the curriculum was focused on purely academic subjects such as philosophy and logic. While Clausewitz personally loved these subjects, he believed that he had been ill served by this course of instruction in his own career. He argued that rather than focusing on the liberal arts, the curriculum should reflect more practical knowledge based on the experiences of the Napoleonic Wars. He also took issue with the method of instruction, which was based largely on rote memorization. Instead, he argued for the case-study method as a means of introducing students to military challenges and guiding them through a process of active learning and problem solving.[8]

Although he had many ideas about how he wanted to alter the course of instruction, he soon found that, as director, he lacked the power to reform the curriculum.[9] To implement his desired reforms, he had to submit all requests to an outside committee that he did not control and was resistant to change. This was infuriating to Clausewitz. Yet again, he had a clear idea of what he wanted to do and an insider's view, but no authority to implement his plans.[10] Even Clausewitz's promotion to major general on September 19, 1818 a significant personal milestone, did not give him greater power or authority over the war college.

In his first few months as director, Clausewitz tried to work within the system to enact his reforms. He drafted numerous proposals for stricter discipline, a more modern curriculum, the inclusion of recent lessons from the Napoleonic Wars, and the addition of a course in formal logic. He submitted his formal request for these reforms in 1819, but it met with little interest from senior officials.[11] Clausewitz's proposal also had the unfortunate timing of coinciding with a political crisis, an economic downturn, and deep cuts in the military budget.[12] Clausewitz's high hopes for transforming the war college were quickly extinguished. Eventually, Clausewitz's proposed changes were adopted in diminished form in 1826 and more fully in the 1860's and the German professional military education system became the finest in the world.[13] However, Clausewitz could not see into the future, and he believed that a perfect opportunity to reform the Prussian military had been lost and that he was being forced to preside over a mediocre and obsolete institution.

Disillusioned, Clausewitz sank into a deep depression. He became passive, often hiding in his office, avoiding contact with students and his fellow professors, and burying himself in reading and writing.[14] His official duties were minimal, typically completed within a few hours every morning and often carried out behind closed doors. His staff brought him official papers to sign at 9:00 a.m. each day, which he often did without speaking a word. He frequently had lunch with Gneisenau and kept up with a surprisingly busy set of social obligations in the evenings, but the majority of his time was spent working on his writings, often in Marie's room or his office.[15] His only interaction with most students and staff came in his role as paymaster for the war college, when he was asked to correct accounting errors or approve salary advances.[16]

By 1823, Clausewitz had become so dispirited with his work at the war college that he agreed to serve conditionally as the Prussian ambassador to London. Although this would have effectively ended his military career, it promised a welcome change of scenery and would have made excellent use of his political skills. He had traveled to London the previous year and possessed strong political instincts, so in certain ways such an appointment made sense. To buttress his case, he penned an essay suggesting that Prussia should adopt the British system of constitutional monarchy, but this was a tone-deaf proposal to King Wilhelm, who already believed that Clausewitz was politically unreliable.[17] Despite this misstep, his appointment had strong support from the chancellor, the foreign minister, and Gneisenau and success seemed assured. With these key endorsements, he was provisionally approved pending the king's signature, but he soon encountered significant resistance from the court. The king refused to endorse Clausewitz's appointment, likely due to lingering resentment over his service in the Russian army a decade earlier.[18] This was a particularly humiliating blow; Clausewitz had not sought out the position but had agreed to serve at the behest of Prussia's leading statesmen, only to be denied by the monarch he had loyally defended. Despite his personal disappointment, Clausewitz did his best to withdraw his name from consideration gracefully, hoping to avoid entanglement in a potentially divisive public conflict between sovereign and state. For a brief period, he was considered for ambassadorships to Switzerland and the German state of Bavaria, but again he was blocked by the king. Although the foreign minister expressed his deepest regrets and held out hope for a future appointment, it was now clear to Clausewitz that he had no future as a diplomat.

Despite his inability to change career paths and his rejection by the monarch, Clausewitz received numerous official rewards and accolades during this period. He was restored to his old post as the military tutor to the crown prince, officially reinstated as a member of the General Staff, and in 1827 his family's claim to nobility was confirmed by royal decree. For most men in their forties who had risen from relatively modest origins, these accomplishments would have been more than enough for a lifetime, yet Clausewitz saw them as insignificant. He wanted nothing less than to achieve lasting fame, but he believed that he had been thwarted by the pettiness of his monarch and that his rightful positions had been taken by less talented men. Clausewitz took these rejections very personally, and for most of his last decade as director of the war college, he was rarely seen around campus, avoided contact with others, and barricaded himself in his office.[19] From the outside, it appeared that Clausewitz was a failure, and several of his fellow faculty members expressed this view in their private writings.[20]

Following these setbacks, the inherently reserved Clausewitz withdrew further from the Prussian social scene and refocused his energies on his academic work.[21] He was exceptionally productive during this period, writing a military biography of Frederick the Great, detailed studies of the 1796, 1799, 1812, and 1815 campaigns, a series of political essays, and his unfinished masterpiece, *On War.*[22] Apart from a brief return to active service in 1830–1831, Clausewitz spent the remainder of his life as a discontented general, a disillusioned director of the war college, and a contemplative philosopher of war.[23]

A Masterpiece Born Out of Revenge

Intensely private and reserved, Clausewitz harbored a deep sense of grievance, first against Napoleon and later against the Prussian elites and contemporary military theorists. In the years that followed, he focused his efforts on the one remaining avenue for asserting his ideas and reputation: writing a major theoretical work on strategy.[24] In this light, *On War* can be understood as the product of his frustration. Ironically, had Clausewitz been appointed ambassador, granted greater military responsibility, or entrusted with more authority at the war college, he would likely have devoted himself to these responsibilities and never taken up his pen to write this book.[25]

For the remainder of his life, Clausewitz was consumed by his writing. While he composed a large number of carefully argued campaign studies, his ultimate objective was much more ambitious. He wanted nothing more than to compose a work that would endure for the ages and "would not be forgotten after two or three years, and that possibly might be picked up more than once by those who are interested in the subject."[26] Clausewitz had read the treatises of contemporary military theorists such as Henry Lloyd, Heinrich von Bülow, and Antonie-Henri Jomini and was profoundly dissatisfied with their efforts.[27] He believed that contemporary military theory was "of limited utility, was displeasing, and lacking entirely in nourishment of the mind." He argued these works lacked the necessary connection with reality and that they are so divorced from reality they "defy common sense." For Clausewitz, "blame lies in the incompleteness of the existing book, and treatments themselves...[as they] lacked the spirit of philosophical inquiry, were usually arranged in a defective manner, with principles and rules being drawn from insignificant bases, and with inconsequential views often being presented as if they were essential."[28] This was a severe and comprehensive critique. As a war college professor and patriot, Clausewitz believed he owed his students and the Prussian nation something better, not "everything obvious that has been stated a hundred times and is generally believed."[29]

While he had been pondering strategic questions for decades, the exact date that Clausewitz began writing *On War* is uncertain. Although the exact starting point is unclear, it was almost certainly after 1815 and possibly several years later.[30] Clausewitz was a reserved perfectionist by nature and seems to have been unwilling to expose his ideas to professional scrutiny until they were fully developed. The book became a solitary endeavor, and Clausewitz shared almost none of his thoughts or details about the book with anyone other than his wife and confidant Marie. Although he had some administrative duties at the war college, these were typically completed by mid-morning, and he used his authority as the director to keep unwanted visitors away and ensure that he had ample time to work uninterrupted.[31]

During this period of isolation, the role of Clausewitz's wife, Marie, was crucial. She was deeply invested in her husband's success and did her best to help him avoid his darker moods and focus on his work. An impressive intellectual in her own right, she often served as a sounding board for her husband's theories and helped solidify and sharpen his logic. Their intellectual and romantic bond was so intimate that much of *On War* was apparently composed in her private bedroom, something that would have been extremely unusual, even socially questionable, at the time. Marie's understanding of her husband's thinking would prove especially important after his unexpected death, as it would fall to her to complete the masterpiece. She passionately believed in the merits of his work and took it upon herself to edit, proofread, and publish her late husband's notes as the book, *On War.*

Recalled to Service

On December 11, 1830, Clausewitz's work on the manuscript was unexpectedly interrupted when he was summoned to Berlin to serve as Gneisenau's chief of staff. Civil war had broken out in Spain and Portugal, and revolutionary unrest was spreading in France, Belgium, Italy, Switzerland, Poland, and several of the German-speaking states. Prussian leaders feared similar unrest and acted preemptively to prevent the revolution from spreading within their borders. Clausewitz was excited to once again be serving in an active capacity and was eager to be reunited with his mentor, even if this necessitated a suspension of his literary efforts.[32]

Before leaving for his new post, Clausewitz carefully organized his incomplete draft of *On War.* He placed the chapters in what he considered a logical order, sealed them into separate packages, and labeled them. He fully intended to return to this "employment [of writing] which he loved so much" and feared that, if left incomplete, his writings could lead to "endless misconceptions." He did not know, however, that he would never return to complete his manuscript.[33]

For a short period, Clausewitz was more content and energetic than he had been in many years. The possibility of French revolutionaries provoking a war with Prussia stirred old animosities, and he could hardly contain his excitement at the prospect of once again experiencing the thrill of battle. Marie, who understood her husband better than anyone else, noted his greatly improved mood. While Clausewitz was somewhat disappointed to be sent east to protect against possible incursions from Poland and Russia rather than west to face France, he was nevertheless happy to be once again putting his martial skills to use.[34] Despite the initial excitement, Clausewitz's last military deployment was largely an anticlimax. In the west, the revolutions of 1830 caused profound changes in the governments of France and Belgium, but they failed to spread into Prussia. To the east, the Polish revolt was rapidly crushed by the Russians, who used it as an excuse to formally annex the rebellious territories, but the rebels enjoyed little sympathy within the borders of Prussia, and the moment of crisis soon passed.[35]

One unfortunate consequence of the revolutions was that the movement of both civilians and troops exacerbated a cholera outbreak that had begun in 1829. By the spring of 1831, just as the revolutions were beginning to ebb, a deadly wave of cholera swept across Europe. This outbreak lasted until 1837 and claimed the lives of hundreds of thousands, possibly millions, across Europe and North Africa. Since they were deployed on Prussia's eastern frontier, Gneisenau and Clausewitz's army were ordered to secure the border with Poland and to cordon off any towns and cities that had been infected. Given the poor understanding of infectious disease and the unsanitary conditions of even the most orderly army camps, this was a dangerous and largely hopeless task.[36]

By early August, Clausewitz believed that the worst of the cholera pandemic was over, and his letters reflected a renewed sense of optimism about the future. This was not to last, however. Clausewitz's mentor and friend, Gneisenau, contracted cholera on August 22, 1831. As Gneisenau's chief of staff and confidant, Clausewitz remained with the dying general until he lapsed into a coma and died the next night. Because he was a cholera victim, Gneisenau was denied a proper military funeral and was immediately cremated. Clausewitz was forced into quarantine after his exposure and was denied the opportunity to pay his respects, a cruel blow that seemed to deepen his sense of loss. [37]

A Sudden Death

While he had seemingly avoided contracting this deadly disease, Clausewitz's letters reveal a dark mood, and his frail constitution appeared to decline after the death of Gneisenau. In September, the final pockets of Polish resistance were crushed, and in early November his army was sent back to its base. He reunited with Marie on November 7, and he looked forward to their intellectual conversations and the opportunity to return to his work *On War*.

He was generally happy over the next few days as he busied himself with myriad social obligations and personal affairs that he had neglected in the previous ten and a half months. On November 16, he worked in his study until lunchtime, when he began to feel unwell. Doctors were called, and they diagnosed him with a mild case of cholera. With rest, he was expected to make a complete recovery, but nine hours later, Clausewitz was dead. His doctors said that his death was not the result of cholera, but that in his fragile physical and mental state that he had simply lost the will to live. Modern scholars have speculated that he died from a heart attack brought on by the stress of recent weeks, but his true cause of death remains unknown.[38] Whatever the cause, Clausewitz died at age 51 without ever achieving the military glory or the intellectual recognition that he so passionately desired.[39]

For the first few decades after his death, it appeared that Clausewitz had died a failure and that his works would be almost totally forgotten. Although *On War* was posthumously published, it sold poorly and was little more than a minor work in the catalogue of military treatises. Indeed, the journey of this book from a minor work to a classic of strategy is a fascinating tale that can help us better appreciate and understand it today.

Notes

1. Parkinson, *Clausewitz: A Biography*, 290.
2. For an overview of the Prussian military reforms, see Craig, *The Politics of the Prussian Army*, Goerlitz, *History of the German General Staff, 1657–1945*; Dupuy, *A Genius for War*; and Citino, *The German Way of War*.
3. Sumida, *Decoding Clausewitz*, 96–97.
4. Parkinson, *Clausewitz: A Biography*, 290–294; and Strachan, *Clausewitz's On War*, 68.
5. Strachan, *Clausewitz's On War*, 68.
6. Paret, *Clausewitz and the State*, 271–272.
7. Strachan, *Clausewitz's On War*, 65.
8. Paret, *Clausewitz and the State*, 274–278.
9. Ironically, Brigadier General Douglas MacArthur faced similar frustrations in 1919 when he attempted to reform the curriculum at West Point. Muth, *Command Culture*, 45–46.
10. Stoker, *Clausewitz: His Life and Work*, 257.
11. During this period, Clausewitz was also involved in discussions regarding the reform of the Prussian militia or *Landwehr*. Clausewitz, *Clausewitz On Small War*, 217–226; and Clausewitz, *Carl von Clausewitz: Historical and Political Writings*, 329–334.
12. Paret, *Clausewitz and the State*, 279.
13. Paret, 279; and Stoker, *Clausewitz: His Life and Work*, 263.
14. Paret, *Clausewitz and the State*, 279–280; and Parkinson, *Clausewitz: A Biography*, 295–297.
15. Bellinger, *Marie von Clausewitz*, 194 and 200.
16. Paret, *Clausewitz and the State*, 307–308.
17. Bassford, *Clausewitz in English*, 37.
18. Paret, *Clausewitz and the State*, 319–323; and Strachan, *Clausewitz's On War*, 64.
19. Parkinson, *Clausewitz: A Biography*, 305.
20. Paret, *Clausewitz and the State*, 325.
21. Paret, 323.
22. Dupuy, *A Genius for War*, 42; Paret, *Clausewitz in His Time*, 77; and Paret, *Clausewitz and the State*, 327 and 330.
23. Parkinson, *Clausewitz: A Biography*, 304–305.
24. Paret, *Clausewitz and the State*, 323.

25. Parkinson, *Clausewitz: A Biography*, 308–309.
26. "On the Genesis of his Early Manuscript on the Theory of War, Written around 1818"; see Clausewitz, *On War* (Howard and Paret), 63.
27. Echevarria, *Clausewitz and Contemporary War*, 13–17.
28. Sumida, *Decoding Clausewitz*, 11. A translation of "Über den Zustand der Theorie der Kriegskunst" appears in Echevarria, *Clausewitz and Contemporary War*, 15–16. For similar statements on the inadequacy of existing military works, see "To an Unpublished Manuscript on the Theory of War, Written between 1816 and 1818" in Clausewitz, *On War* (Howard and Paret), 61–62.
29. "On the Genesis of his Early Manuscript on the Theory of War, Written around 1818" ; see Clausewitz, *On War* (Howard and Paret), 63.
30. Echevarria, *Clausewitz and Contemporary War*, 4.
31. Martin van Creveld argues that the absence of external pressure to complete *On War* was critical to its eventual success, as it allowed Clausewitz the intellectual freedom and time to produce an enduring work. See van Creveld, "The Eternal Clausewitz," 48.
32. Aron, *Clausewitz: Philosopher of War*, 38.
33. Paret, *Clausewitz and the State*, 326; and Parkinson, *Clausewitz: A Biography*, 320.
34. Parkinson, *Clausewitz: A Biography*, 321–322.
35. As part of his preparations for operations in Poland, Clausewitz conducted a detailed study of the 1793 and 1794 Polish campaigns. See Stoker, *Clausewitz: His Life and Work*, 278–279; and Lebow, *The Tragic Vision of Politics*, 214.
36. Paret, *Clausewitz and the State*, 421–423.
37. Paret, 425–427; and Parkinson, *Clausewitz: A Biography*, 327–328.
38. Paret, *Clausewitz and the State*, 429–430; Parkinson, *Clausewitz: A Biography*, 329–330; and Strachan, *Clausewitz's On War*, 67.
39. Paret, *Clausewitz and the State*, 432.

CHAPTER 6

FROM FAILURE TO CANON: THE POSTHUMOUS JOURNEY OF *ON WAR*

Clausewitz's awareness of his mortality had long shaped his thinking about *On War*. Nearly four years before his death, he suffered one of his recurring illnesses and worried that his masterpiece would remain unfinished. He viewed the work as his greatest legacy but admitted, "Should I be interrupted in this work by an untimely death, what exists of it may certainly be described as merely a hotchpotch of ideas, which being exposed to ceaseless misunderstandings, will give rise to a multitude of criticisms."[1] His concerns were not misplaced. He died before he could complete *On War*, and for nearly two centuries, his work has been the subject of ongoing criticism and misinterpretation.[2]

Marie von Clausewitz
Without the extraordinary efforts of Marie von Clausewitz, Clausewitz's name and works might well have been lost to history. Marie was a phenomenal woman with education, social graces, connections, and a

burning desire to promote her husband and his ideas.[3] From their earliest meetings and courtship, the two treated one another as intellectual equals and had enjoyed the thrust and parry of academic debates. Because of the social mores of the period, much of this relationship was conducted in private, but Marie could rightly claim, as she did in her preface to *On War* that "we shared everything."[4] Indeed, there is good reason to believe that Marie read, understood, and even helped shape her husband's works, softening some of his views and language during the composition process.[5] During his life, Clausewitz would often joke to Marie that he hoped she would be able to complete his work in the event of his untimely death. Though she may not have taken his dark humor seriously at the time, Marie quickly assumed the role Clausewitz had prophesied in jest.[6]

Almost immediately after her husband's death, Marie sprang into action.[7] Just two days after his passing, she wrote to her friend Countess Bernstorff, expressing her grief and beginning to shape the narrative of Clausewitz as a tortured genius:

> It was a great comfort to me that at least his final moments were calm and painless, and yet the expression and tone with which he expelled his last sigh were heart rendering; it was though he pushed life away like a heavy burden...indeed life for him had consisted of an almost unbroken chain of effort, sorrow, and vexation. Certainty, on the whole he had achieved much more than he could have hoped for at the outset; he felt this deeply and acknowledged it with a grateful heart. But he never did scale the highest peak, and every pleasure that he experienced contained a flaw that clouded his enjoyment. He lived through great, glorious years, but never had the good fortune of fighting in a victorious battle...To a rare degree he enjoyed the friendship of the most noble men of his time, but he never received the recognition that alone would have enabled him to be genuinely useful to his country. And how did he not suffer with and for his friends.[8]

Marie saw her husband's unfinished manuscript as his final and best opportunity to secure the recognition he had been denied in life. She was correct—the book would ultimately become one of the most venerated texts in the strategic cannon, though achieving such lasting fame required significant effort, favorable circumstances, and the passage of time.

Within days of Clausewitz's death, Marie decided to publish his unfinished manuscript. As she opened the envelopes that he had sealed prior to his recall to active duty in 1830, she likely felt the profound weight of responsibility. She soon devised a plan to publish a ten-volume edition of Clausewitz's writings, with the first three volumes comprising *On War* and the final seven presenting his collected lectures, notes, and papers.[9] The responsibility of editing, organizing, publishing, and promoting this material now rested with her—a task made urgent while the memory of Clausewitz remained vivid and his professional network remained engaged. Fortunately, Marie could count on the support of her brother, Lieutenant General Fredrich Wilhelm von Brühl, and two of her husband's close friends, Carl von Gröben and Major Franz August O'Etzel.[10] Although the precise extent of their contributions is unclear, it is generally accepted that Brühl assisted in arranging the chapters and negotiating with publishers, while O'Etzel contributed maps for one of the early editions.[11]

While many questions about the editing process will likely never be answered, it is certain that Marie played an active and decisive role.[12] She was protective of her husband's work and maintained strict physical control over the manuscript pages. Her editorial approach appears to have favored minimal revision.[13] Aware that the work was incomplete and contained contradictions, she nonetheless believed that it was best to preserve the text largely in its original form, allowing her husband's intellectual voice to remain intact.[14] This proved to be a prudent decision: modern scholars benefit from a relatively unaltered version of the manuscript, and the ambiguity in Clausewitz's writings has fostered ongoing scholarly debate and interpretation.[15]

Marie had much to take pride in for her efforts. She preserved her husband's works, oversaw their publication, resisted pressures to implement major posthumous revisions, distributed copies to members of Prussian society and the royal family, and established the narrative of Clausewitz as a misunderstood genius. Although she died in 1836 at the age of fifty-six, she initiated a legacy that gradually fulfilled what she had always hoped for—lasting recognition and fame for her husband, which had eluded him during his lifetime.[16]

Subsequent editors and commentators, however, have not always demonstrated the same restraint and or respect for Clausewitz's original intent as Marie had. Shortly after Marie's death, her brother Fritz von Brühl introduced several hundred changes to the manuscript, which he published in 1853 as a revised second edition. This version became the basis for most translations until after World War II and contributed significantly to the persistent misunderstandings of Clausewitz's ideas.[17]

Slow Recognition of Clausewitz's Legacy

Despite Marie's commitment to preserving and promoting her husband's work, *On War* was not an immediate commercial success. Rather than providing the financial and reputational rewards she had hoped for, the book languished, selling fewer than one hundred copies in the decade following its publication. Although reviewed in some military journals, the text remained a niche work with limited appeal. This reception was largely due to the book's character as a dense philosophical treatise rather than a practical military manual. As one contemporary German review aptly observed, the book, "cannot be read it has to be studied."[18]

Within the Prussian army, Clausewitz's work appears to have had minimal influence in the decades leading up to the Wars of German Unification. *On War* was not part of the curriculum at German military academies, and most senior military leaders made little or no reference to the work in their writings.[19] Helmuth von Moltke would later claim to have drawn significant inspiration from the text, but the statesman Otto von Bismarck admitted that he had never read it.[20] Clausewitz is therefore best understood as one among many military theorists of the period, whose work was neither widely read nor institutionally embraced until well into the twentieth century. In fact, a sustained theoretical engagement with Clausewitz's ideas does not appear to have become either a requirement or a priority for aspiring German officers until after World War II.

To the extent that Clausewitz and his ideas were used, they often appeared in a distorted or superficial forms. In the decades preceding World War I, for instance, his concept of "absolute war" was appropriated and politicized by German militarists advocating for larger military budgets and aggressive strategies. General Alfred von Schlieffen, architect of the infamous Schlieffen Plan for the invasion of Belgium and France, scattered references to *On War* throughout his writings. However, despite these allusions, Schlieffen fundamentally misunderstood several of Clausewitz's central concepts, including friction, the political nature of war, the center of gravity, the culminating point of victory, and the role of human factors. These conceptual misreadings undermined the strategic soundness of his plan.[21] Likewise, Clausewitz's concepts of military genius and moral forces in warfare were sometimes used to resist the modernization of the German military, privileging leadership and *esprit de corps* over technological advancement.[22] In the most extreme case, General Friedrich von Bernhardi distorted Clausewitz's famous dictum that war is a continuation of politics by other means, arguing that the ideal general should ignore and overcome political constraints rather than incorporate them into strategic calculations.[23]

A common theme in these misappropriations of Clausewitz was their use as post hoc rationalizations for policy decisions that had already been made. That these arguments often bore little resemblance to Clausewitz's actual theories mattered little. The mere invocation of his name lent authority and gravitas to arguments advanced before audiences largely unfamiliar with his work. This superficial engagement with Clausewitz's ideas suggests that his writings were not deeply studied within the

German military prior to World War I, and that to the extent they were referenced, they often appeared in distorted or caricatured form.

The Slow Proliferation of *On War*

Outside of Germany, *On War* achieved acceptance even more slowly, particularly in the English-speaking world. The first English translation by J.J. Graham, a colonel in the British Army and a serious student of military theory, was not published until 1873. While this translation made Clausewitz accessible to English-speaking audiences, it was deeply problematic. Hastily produced in the wake of the Franco-Prussian War, which had generated interest in understanding the sources of German military power, the translation suffered from significant flaws. This edition was seriously flawed because Graham based his work on the third German-language version, which had already incorporated numerous editorial alterations. This choice compromised the accuracy of the translation and set a flawed precedent that would persist in English editions for more than a century.[24]

If the flawed source text were not problematic enough, Graham's limitations as a linguist further compounded the issue. Clausewitz wrote in a dense, academic, and somewhat archaic form of German that remains challenging even for native speakers and resists direct translation into English.[25] His prose is marked by frequent use of the passive voice, intricate sentence constructions, and multiple dependent clauses, all of which complicate both comprehension and translation.[26] Confronted with such challenging material, Graham produced a translation that was overly literal and failed to capture many of the nuances in Clausewitz's original language. Although subsequent English editions appeared over the next century, including Otto Jolles's generally respected 1943 version, Graham's translation nonetheless became the dominant English-language edition.[27] The result was that generations of readers and scholars engaged with an incomplete and linguistically flawed rendering of an already difficult text.

Perhaps no English-language edition of *On War* has generated as much confusion and misunderstanding as the 1968 abridgement by Anatol Rapoport.[28] Rapoport was a complex and intellectually versatile figure. Born in Russia in 1911 into a secular Jewish family, which emigrated in 1922 to the United States, where he was exposed to a wide array of educational and cultural opportunities. A musical prodigy, he mastered the piano and music theory and pursued conducting, eventually traveling to Vienna for further study. The rise of fascism and growing antisemitism, however, compelled him to return to the United States in 1934, at which point he abandoned music and turned to mathematics, earning a PhD from the University of Chicago. Though he briefly engaged with communist politics, he later served in the U.S. Army during World War II. After the war, he resumed his academic pursuits in Chicago and became a pioneer in the emerging field of mathematical biology. He went on to hold positions at the University of Michigan and the University of Toronto. Over the course of his long and distinguished career, Rapoport worked across multiple disciplines—including mathematics, psychology, and peace research—and came to exemplify the modern polymath.

Despite his impressive intellectual credentials, Rapoport's edition of *On War* left much to be desired. He based his abridgement on the 1873 Graham translation which—despite its known deficiencies—remained the standard English edition at the time.[29] Building on this problematic foundation, Rapoport proceeded to excise substantial portions of the text, including the entirety of Books V, VI, and VII. While these sections do contain elements specific to Napoleonic-era military practices, they also include vital discussions of limited war, the difficulty of conquest, and the primacy of political aims. Though the abridgment significantly reduced the length of the book, it did so at the cost of removing many of Clausewitz's most important reflections on the necessity of moderating violence in war.

In his role as editor, Rapoport also composed an introductory essay in which he was sharply critical of Clausewitz and his work. He argued that the modern nation-state system was obsolete, that violence was ultimately futile, and that war should no longer be regarded as a legitimate instrument of statecraft. While Rapoport acknowledged Clausewitz's insightful analysis of Napoleonic warfare, he warned that the "neo-Clausewitzian" application of these theories was dangerous and immoral:

> Napoleon taught one lesson: the universal currency of politics is power, and power resides in the ability to wreak physical destruction. Clausewitz embodied this lesson in unifying a philosophy of politics with a philosophy of war.[30]

Given that Rapoport had personally witnessed half a century of war and genocide, his deeply held convictions about modern conflict are understandable. Nevertheless, his edition did a considerable disservice to the interpretation of Clausewitz by reinforcing the misconception that *On War* represents the outdated and reductive thinking of a warmonger.[31] Although Rapoport's abridgment has been widely criticized by Clausewitz scholars, it remains in print as a low-cost paperback and continues to mislead new generations of students and practitioners.

Jomini as a Dynamic Foil for Clausewitz

While Clausewitz's *On War* languished for over a century due to poor sales and flawed translations, *The Art of War* by Henri Jomini flourished in the decades following the Napoleonic Wars. Although both men served as staff officers during the Napoleonic era and authored enduring military treatises, they were markedly different in temperament and style.[32] Unlike the reserved and introspective Clausewitz, Jomini was an assertive self-promoter who almost certainly exaggerated his proximity to Napoleon and often boasted of his ability to anticipate the emperor's decisions. While Clausewitz was risk adverse and discreet, Jomini's insecurities found expression in overconfidence and in sharp criticism of those he perceived as rivals.[33] He was particularly hostile toward Clausewitz, launching *ad hominem* attacks on his writing and dismissing his logic as "frequently defective."[34] Jomini also enjoyed considerable acclaim during his own lifetime. Unlike Clausewitz, who died before *On War* was published, Jomini lived for decades after the release of *The Art of War*, actively revising, promoting, and defending his ideas. Indeed, for much of the next century, Jomini's work proved far more influential among military thinkers.

There are several reasons for Jomini's early dominance. Unlike Clausewitz, Jomini did not aim to produce a philosophy of war. Instead, he wrote a work designed for broad appeal.[35] *The Art of War* was more tactical and operational in focus, and its style was simpler and less abstract. For readers seeking practical solutions, Jomini's more accessible format offered a quicker and easier alternative to Clausewitz's dense and intellectually demanding tome. Moreover, Jomini's work was written in French, a language far more widely read and spoken than German at the time, particularly within the American military. This linguistic advantage, combined with Jomini's prolific coinage of military terms, enabled him to shape the professional lexicon and effectively define the terms of military discourse.[36] The use of French military terms was so ubiquitous that West Point required cadets to study French, as it had become the military and scientific lingua franca of the period due in large part to the widespread adoption of Jominian terminology. This dominance of Jominian terms was evident during the American Civil War, as officers routinely employed phrases such as *attack in echelon, oblique movement, coup de main,* and other military jargon.[37]

Jomini also benefited from the fact that his theory resembled a checklist or set of best practices, rather than an abstract theory of war. He claimed that his work was grounded in "fundamental, immutable principles."[38] As he boasted:

> Correct theories, founded upon right principles, sustained by actual events of wars, and added to accurate military history, will form a true school of instruction for generals. If these means do not produce great men, they will at least produce generals of sufficient skill to take rank after the natural masters of the art of war.

Jomini went so far as to claim that adherence to his methods would guarantee victory, stating, "I have not found a single case where these principles, correctly applied did not lead to success."[39]

Such an approach reflects a profound epistemological confidence, yet it was consistent with the broader nineteenth-century drive toward rationalism in scientific and military thought. If complex problems could be broken down into smaller components and analyzed systematically, then a more complete understanding—and ultimately mastery of—the world seemed possible. Indeed, during this period, significant advances in medicine, transportation, manufacturing, communication, and chemistry were achieved using this very approach. Why, then, should war and strategy be any different?

Compared to Clausewitz's refusal to offer prescriptive formulas for success, Jomini's promise of a rational, systematic path to victory held considerable appeal.[40] In the conclusion of his work, Jomini even goads the reader with an explicit assurance of triumph:

> If a few prejudiced military men, after reading this book and carefully studying the detailed and correct history of the campaigns of the great masters of the art of war, still contend that it has neither principles nor rules, I can only pity them, and reply, in the famous words of Frederick, that 'a mule which had made twenty campaigns under Prince Eugene would not be a better tactician than at the beginning.'[41]

Relative to Clausewitz's more abstract and internally inconsistent text, the appeal of a promised scientific method for achieving victory proved irresistible.[42]

While Jomini's *Art of War* is largely forgotten today, it is important to remember that it once held hegemonic status in terms of both commercial success and strategic influence. The American military, in particular, enthusiastically embraced his approach; it was said—only half in jest—

that officers during the American Civil War went into battle with, "a sword in one hand, and Jomini in the other."[43]

The Enduring Wisdom of Clausewitz's Approach

In contrast to Jomini's prescriptive, checklist-style approach, Clausewitz fundamentally rejected the notion that warfare could be reduced to a set of scientific principles or universal best practices. With Jomini clearly in mind, Clausewitz lamented:

> Efforts were therefore made to equip the conduct of war with principles, rules or even systems. This did present a positive goal, but people failed to take adequate account of endless complexities involved. As we have seen, the conduct of war branches out in almost all directions and has no definite limits; while any system, any model, has the finite nature of a synthesis. An irreconcilable conflict exists between this type of theory and actual practice.[44]

Clausewitz continued his critique of Jomini, even subtly referencing the title of Jomini's most famous work, *The Art of War*, in his dismissal: "to reduce the whole secret of the art of war to the formula of numerical superiority *at a certain time in a certain place* was an over-simplification that would not have stood up to the realities of life."[45]

Rather than offering a series of instructions for winning battles, Clausewitz envisioned his theories as a means of cultivating the strategist's intellect. Under the heading "Theory Should Be Study, Not Doctrine," he articulated the proper role of theory—as a guide for critical thinking rather than a prescriptive manual, noting:

> [theory should not be] a positive manual for action...Theory will have fulfilled its main task when it is used to analyze the constituent elements of war, to distinguish precisely what at first sight seems fused, to explain in full the properties of the means employed and to show their probable effects, to define clearly the nature of the ends in view, and to illuminate all phases of warfare...[Theory] is meant to educate the mind of the future commander, or, more

accurately, to guide him in his self-education, not to accompany
him to the battlefield; just as a wise teacher guides and stimulates
a young man's intellectual development, but is careful not to lead
him by the hand for the rest of his life.[46]

Clausewitz believed that theoretical works should shape thinking around
general principles and stimulate further reflection. Jomini's approach was
foolish and dangerous because it risked lulling readers into a false sense
of confidence and impeding critical thinking rather than encouraging it.

While Clausewitz acknowledged that his more philosophical approach
to war was more challenging, he ultimately believed it offered the most
effective means of understanding such a complex phenomenon:

> Philosophy teaches us to recognize the relations that essential
> elements bear to one another, and it would indeed be rash from this
> to deduce universal laws governing every single case, regardless of
> all haphazard influences. Those people, however, who *never rise*
> *above anecdote'* as a great writer said, and who would construct all
> history of individual cases—starting always with the most striking
> feature, the high point of the event, and digging only as deep as
> suits them, never get down to the general factors that govern the
> matter. Consequently their findings will never be valid for more
> than a single case; indeed they will consider a philosophy that
> encompasses the general run of cases as a mere dream (emphasis
> in the original).[47]

In embracing this complex and philosophical approach, Clausewitz
demanded a great deal from his readers. Despite the richness of his
theoretical insights, few have taken the time to fully engage with the
intricacies of his ideas. It is therefore no surprise that *On War* was long
regarded as little more than a minor work, overshadowed by simpler
and less theoretical treatises on strategy.

Rediscovering Clausewitz in America: A Four-Part History
"Why was Clausewitz forgotten?" The answer is straightforward: *On*
War was a complex philosophical treatise characterized by dense prose,

poor sales, and flawed translations, making it easy to overlook. A more compelling question might be, "why was Clausewitz rediscovered?" In the American context, the answer is surprising and stems from an unlikely combination of factors: the advent of nuclear weapons, a renewed emphasis on professional military education, the publication of a new translation, and the American defeat in the Vietnam War. Together, these developments transformed *On War* from a neglected specialist text into an unexpected bestseller and cultural touchstone.

Nuclear Weapons

As the Americans and British raced to develop the atomic bomb, their overriding objective was speed. They feared that Nazi scientists might win the atomic arms race and plunge the world into what Winston Churchill described as, "new Dark Age made more sinister, and perhaps more protracted, by the lights of perverted science."[48] Faced with such a doomsday scenario, speed took precedence over humanitarian concerns, fiscal constraints, or the long-term implications of atomic weapons. Even after Germany's defeat, the war against Japan continued, and development of the bomb proceeded. Once these weapons became available, there was little delay in employing them. Two atomic bombs were dropped on Japan and the most destructive war in human history ended abruptly. America now held a nuclear monopoly but had done little thinking about how these weapons should be incorporated into its strategy.[49] Had war been fundamentally transformed? Was war now impossible? Was global destruction a real possibility? This was uncharted territory, and no one truly knew the answers.

One of the first people to think seriously about the implications of nuclear weapons was Bernard Brodie, an American military strategist, naval historian, and professor at Yale University.[50] On paper, Brodie was not particularly qualified to weigh in on nuclear strategy. Nevertheless, he recognized the subject's vital importance and did not wait for others to take the lead. Drawing on his expertise in military history and theory, he applied these insights to the unprecedented challenges posed by nuclear weapons. Because of his academic background, Brodie was familiar with Clausewitz's work—at a time when few others were—and he quickly adapted the Prussian theorist's ideas to his purposes. Faced with the prospect of nuclear war, Clausewitzian insights took on new urgency: that violence tends toward extremes, that war cannot be won in a single blow, and that war must serve a political purpose. Brodie now had his theoretical framework.[51]

Brodie's 1946 book, *The Absolute Weapon*, developed these insights into a groundbreaking work of strategy. The book became an instant classic as it effectively defined the terms of debate on nuclear weapons and deterrence theory. According to Brodie, the unprecedented destructive power of nuclear weapons fundamentally altered the political calculus of warfare. In a nuclear age, violence could no longer be taken to extremes, wars could not be won, and achieving political objectives through military means would be nearly impossible without incurring unacceptable costs.[52] Although Brodie never mentioned Clausewitz by name in this work and tended to oversimplify his ideas, his central argument—that nuclear Armageddon rendered total war both politically impossible and militarily unwinnable—was both logically persuasive and ethically resonant.[53]

By moving first and articulating a compelling argument, Brodie quickly emerged as a leading voice in the nascent debates on nuclear strategy and ethics. In later books and articles, Brodie explicitly cited Clausewitz's ideas as the theoretical basis of his work and soon earned the moniker "the American Clausewitz." Thanks to Brodie's influence, a basic familiarity with Clausewitz's work became essential for engaging in contemporary discussions about nuclear weapons.[54] Although no suitable English translation of *On War* existed at the time, during the early decades of the Cold War, a range of prominent authors—including Herman Kahn, Henry Kissinger, Robert Osgood, Thomas Schelling, and others—invoked Clausewitz as an intellectual foundation for their theories of nuclear conflict.[55] While this was not sufficient to propel Clausewitz's ideas into the mainstream, it marked a critical turning point.[56]

Reemphasis of Professional Military Education

While Clausewitz's *On War* had been studied in the American military for decades, the effort was often haphazard. His work appeared intermittently on the syllabi of various officer candidate schools, military academies, staff colleges, and war colleges. However, following World War II, interest in the Napoleonic Wars in general and the Prussian theorist in particular had significantly waned. Clausewitz's eventual rise to prominence within the American military education system resulted from two unexpected developments: the 1968 Tet Offensive and the initiative of Lieutenant Colonel David MacIsaac, an instructor at the United States Air Force Academy.[57]

In the immediate aftermath of the Tet Offensive, MacIsaac recognized that the political narrative of the Vietnam War had fundamentally shifted. Seeking a framework to understand and discuss the political underpinnings of strategy, he successfully advocated for a "translated and distilled" version of *On War* to be assigned to cadets during the 1968–1969 academic year. Encouraged by the positive response from his cadets, MacIsaac lobbied for more time to study the Prussian theorist, a request that was approved by his academic supervisors. Thanks to these efforts, Clausewitz's work began to gain a deeper foothold in the American military education system. Yet, MacIsaac was far from finished. In 1975, he took a one-year sabbatical to teach at the Naval War College and brought *On War* with him. Under the direction of Admiral Stansfield Turner, the Naval War College was already engaged in a major curriculum overhaul, emphasizing classic texts and greater academic rigor. MacIsaac found a receptive audience, and a comprehensive study of *On War* became a permanent feature of the curriculum. In 1979 MacIsaac was reassigned to the Air War College, where he again succeeded in integrating Clausewitz's work into the curriculum. Finally in 1981, through continued lobbying from afar, he convinced the Army War College to adopt *On War* into its program of study.[58]

While David MacIsaac played a key role in planting the seeds of Clausewitzian thought throughout the American military education system, fundamentals changes were beginning to take place in the American military. Much like the Prussians after their defeat at Jena and Auerstädt, the United States responded to its defeat in Vietnam by fundamentally rethinking its approach to war.[59] This reassessment led to a wide range of reforms, including the transition to an all-volunteer force, the development of a new generation of weapons systems, an emphasis on more realistic training, a heightened awareness of the limits of American power, a reexamination of the civil-military relationship, and an increased emphasis on professional military education. Education was a central component of this multifaceted reform program. Vocational training, college funding, and the pursuit of advanced degrees for officers were all part of a broad-based effort to improve the military's human capital. Although some questioned this approach, the investment quickly yielded results in the form of higher-quality recruits, increased professionalism, better retention, and enhanced combat effectiveness.

In this new environment, Clausewitz's *On War*—even in its abridged and imperfectly translated form—quickly became required reading for aspiring officers. Almost overnight, it joined other foundational works such as Thucydides' *History of the Peloponnesian War* and Sun Tzu's *Art of War* as integral components of the revamped curricula at the various staff colleges and war colleges.[60] These "classic" texts were ideally suited for a military seeking to enhance its professional credibility and reevaluate its strategic orientation. Rich in theory and intellectually demanding, they also offered a more indirect means of confronting American military shortcomings. While the raw experiences of Khe Sahn or My Lai might be too personal for frank discussion, officers could more comfortably examine the failures of Alcibiades or Napoleon and extract valuable lessons from history. Building on these early successes, works such as *On War* came to serve as a rite of passage in the education of military professionals.

The experience of future chairman of the Joint Chiefs of Staff Colin Powell illustrates the military's new emphasis on classic texts. In 1975, when Powell, then a lieutenant colonel, enrolled at the National War College, he had little exposure to strategic theory in general or Clausewitz's work in particular.[61] Prior to attending, he had seriously considered leaving the army but chose to stay for the educational benefits—and he flourished. He found genuine inspiration in *On War*, which he described as a "beam of light from the past, still illuminating present day military quandaries." Through Clausewitz's lens, the traumatic experiences of Vietnam suddenly began to make sense as:

> Clausewitz's greatest lesson for my profession was that the soldier, for all his patriotism, valor, and skill, forms just one leg in a triad. Without all three legs engaged, the military, the government, and the people, the enterprise cannot stand.[62]

Based in no small part on the perspectives he gained from engaging with Clausewitz's ideas, Powell was able to make sense of the Vietnam War, regain a sense of purpose and optimism, and rededicate himself to his profession.

Powell's unique combination of talents enabled him to serve as a White House Fellow, deputy national security advisor, assistant to Secretary of Defense Casper Weinberger, national security advisor, and chairman of the Joint Chiefs of Staff. One of Powell's most enduring legacies as chairman was the development of the Weinberger-Powell Doctrine, which was informed by his engagement with Clausewitz's work. The doctrine emphasized the necessity of clear objectives, overwhelming force, public support, and a defined exit strategy as prerequisites for the use of American military power.[63] Clausewitz himself would have likely been dismayed by such a rigid and formulaic interpretation of his theories, but the doctrine nonetheless represented a sincere attempt to apply his emphasis on aligning ends and means and on recognizing the political limits to American foreign policy.[64] In the aftermath of the Vietnam War, this clarity of purpose and insistence on decisive outcomes proved especially appealing. The swift victory in Operation Desert Storm appeared to vindicate this (albeit debatably) "Clausewitzian" approach.

Another example of the American military's embrace of Clausewitzian thinking was the 1989 Marine Corps doctrinal publication, *Fleet Marine Force Manual 1: Warfighting* (FMFM-1). This foundational doctrine was developed by a group of "warrior-scholars" who had firsthand experience with the complexity and chaos of combat and had studied *On War* during their time in the professional military education system. With the support of General Alfred M. Gray Jr., then commandant of the Marine Corps, they echoed Clausewitz's observations that war is an inherently human endeavor marked by uncertainty, friction, and constant change. Rather than offering a checklist of best practices, FMFM-1 embraced complexity, acknowledged the emotional and psychological dimensions of war, and emphasized the necessity for adaptive leadership.[65] This Clausewitzian philosophy continues to shape the Marine Corps' approach to warfighting and remains its central doctrinal text.

The 1976 Howard and Paret Translation

Although Clausewitz's theories were now becoming a major part of nuclear debates and professional military education, the English-speaking world still lacked a complete and reliable translation of *On War*. If *On War* was ever to be something more than a text for specialists, this deficiency needed to be addressed.[66] The first steps toward remedying the problem were undertaken by the German American academic Peter Paret, who founded the Clausewitz Project at Princeton University in 1962.[67] Paret's influence on the modern understanding of Clausewitz's theories should not be underestimated. As a specialist in nineteenth-century Prussia, he correctly recognized Clausewitz as a key intellectual figure and dedicated much of his career to popularizing his work.

To fund his research, Paret obtained several US government grants specifically for the purpose of translating and publishing a new edition of *On War*—a clear indication that the military was invested in making the text more accessible—and soon dedicated his life to the study and popularization of Clausewitz's theories. Paret used this funding and his academic connections to host conferences, bring together scholars, and generate interest in Clausewitz's work. Through these efforts, Paret enlisted two eminent academics, Bernard Brodie and Michael Howard, to assist him in the monumental task of producing a new translation of *On War*. Brodie served as an editor and contributed introductory essays, while Howard worked as a co-translator and contributed several essays to the edition.

Howard and Paret labored on the translation project for over a decade. In the process, they corrected numerous errors and omissions, improved the consistency of the language, and significantly enhanced the overall readability of the text. In addition, they included a series of essays that clarified key elements of Clausewitz's theories, provided historical context, and addressed longstanding misconceptions about his work. Although their translation generated some minor controversy over specific word choices and was criticized for its inadequate index, it was a major scholarly achievement.[68] This edition surpassed both the Graham translation and the Rapoport abridgement and remains the authoritative version for soldiers and scholars alike.

Based on a combination of academic, military, and general-interest purchases, the Howard and Paret translation quickly became the best-selling edition of *On War* of all time, far exceeding the success of the original German-language version.[69] While the Howard and Paret translation was a critical breakthrough, it did not end the misunderstanding and misuse of Clausewitz. Even in this more digestible form, the text has remained extremely dense, and many of the prior misconceptions about the work have endured.

Clausewitz and the Vietnam War

The defeat of the United States in the Vietnam War was a watershed moment for America. It was a shocking realization that the richest and most powerful nation on Earth could be thwarted by a relatively impoverished and weak adversary. One of the unexpected consequences of this traumatic experience—alongside the era's cultural upheavals—was a renewed appreciation of Clausewitz's *On War*. While Clausewitz's work had already seen a resurgence among intellectuals and within professional military circles, his theories became more mainstream in the decades following the defeat in Vietnam.[70] His insights on friction, the limits of intelligence, the political nature of war, the trinity, the difficulty of achieving victory through offense, and the power of nationalism to fuel people's wars offered a compelling framework for understanding the US failure in the Vietnam War.

In the wake of the Vietnam War, a growing body of strategists turned to Clausewitz's work to make sense of the American defeat. Among them, the most influential was Colonel Harry G. Summers, whose 1982 book *On Strategy: A Critical Analysis of the Vietnam War*, became a foundational text in military education. Summers, an infantry officer who had served in combat in Korea and Vietnam, was widely respected for his personal courage and intellectual insight. Summers began his analysis by posing a painful question, "How could we have succeeded so well [on the battlefield], yet failed so miserably?"[71] He contended that America lost the Vietnam War because its leaders fundamentally misunderstood Clausewitzian principles. By failing to declare war and articulate clear political objectives, American leaders lacked a coherent understanding of what they were trying to achieve. This failure was compounded by a disregard for friction, as planners routinely assumed conditions were more favorable than they were. Moreover, the elements of Clausewitz's trinity —government, military, and people—were out of alignment. The American public and Congress were unwilling to support the war effort fully and often undermined military operations.[72] As Summers bitterly remarked:

> By default the military had allowed strategy to be dominated by civilian analysts—political scientists in academia and systems

analysts in the defense bureaucracy...In justifying strategy in civilian terms, the Army surrendered its unique authority based on battlefield experience.[73]

The implication was clear: the Vietnam War might have been winnable if the elements of Clausewitz's trinity had been better aligned and if the military had been granted greater autonomy in formulating and executing policy.[74]

In the decades following the Vietnam War, Summers's work exerted considerable influence within the American military, though this influence was something of a mixed blessing.[75] Its entire premise was based on an overly simplified version of Clausewitz's work in general and his theory of the trinity in particular. Summers omitted the original formulation of the trinity (primordial violence, hatred, and enmity), overemphasized the rationalist elements of Clausewitz's theories, and underrepresented the political divisions and social ferment of the Vietnam Era.[76]

Although it was based on an imperfect understanding of both Clausewitz's theories and the historical evidence, Summers's invocation of *On War* inspired many imitators. Borrowing from his approach, dozens of books published over the next decade used Clausewitzian concepts as their theoretical foundation. Soon, references to Clausewitz's work became ubiquitous among armchair strategists attempting to analyze a wide range of wars and military campaigns. This trend was problematic, however, as the application of his complex theories was often superficial, overly deterministic, or misinterpreted.[77] Even so, in the years following the American defeat in the Vietnam War, *On War* entered the mainstream discourse, due in large part to a spate of books seeking to rationalize that traumatic failure and to a wave of authors who extended this analytical framework to other conflicts.

To avoid falling into this same trap, the following chapters will offer guidance on and analysis of several of Clausewitz's most important—and frequently misunderstood—theories.

Notes

1. Strachan, *Clausewitz's On War*, 73
2. Raymond Aron speculated that Clausewitz was too proud and too sensitive to criticism to have published such a work during his lifetime. Aron, *Clausewitz: Philosopher of War*, 1.
3. Bellinger, *Marie von Clausewitz*; and Bellinger, "The Other Clausewitz," 345–367.
4. Bellinger, *Marie von Clausewitz*, 197.
5. Bellinger, 192–196.
6. Bellinger, 195; and Stoker, *Clausewitz: His Life and Work*, 281.
7. Bellinger, *Marie von Clausewitz*, 218–219.
8. Bellinger, *Marie von Clausewitz*, 219–221 and Paret, *Clausewitz and the State*, 431.
9. Bellinger, *Marie von Clausewitz*, 224.
10. Bellinger, 223.
11. Friedrich Wilhelm von Brühl contributed to the misunderstanding of Clausewitz's views on limited war through his edits to the second German edition, particularly in Book VIII, Chapter 6. These changes distort Clausewitz's theory and make it appear more militaristic than originally intended. Strachan, *Clausewitz's On War*, 69.
12. Bellinger, *Marie von Clausewitz*, 223–225. For more critical views, see Gat, *A History of Military Thought*, and Sumida, *Decoding Clausewitz*.
13. The other campaign studies appear to have been more heavily edited by O'Etzel and Gröben. Bellinger, *Marie von Clausewitz*, 224.
14. O'Etzel and Gröben apparently agreed with this editorial decision as there is no evidence of their opposition or disagreement in their correspondence. Bellinger, *Marie von Clausewitz*, 224–225 and 230.
15. Interestingly, the first criticism of Marie's editorial efforts appeared in 1833, the year after the work was published; Bellinger, *Marie von Clausewitz*, 226.
16. Bellinger, *Marie von Clausewitz*, 232–235.
17. Bellinger, 225–226.
18. Muth, *Command Culture*, 20.
19. Muth, 21.
20. Ironically, Bismarck appears to have implicitly understood the political nature of war better than Moltke, despite never having read Clausewitz's

work. Craig, *The Politics of the Prussian Army*, 181. Moltke had attended the war college during Clausewitz's tenure as director, but there is no record of their interacting, as Clausewitz did not teach during this period and lived in a self-imposed isolation. Dupuy, *A Genius for War*, 61 and 107; and Muth, *Command Culture*, 20.

21. Chickering, *Imperial Germany and the Great War, 1914–1918*, 20. See also Foley, *German Strategy and the Path to Verdun*, 14, 40–41, and 46.
22. For a detailed discussion of the internal tensions within the German Imperial Army regarding technology adoption, see Brose, *The Kaiser's Army*.
23. Bernhardi, *Germany and the Next War*, 35. See also Echevarria, *After Clausewitz*, 7, 14–15, 110–113, 190, 206, 221, 225; Heuser, *Reading Clausewitz*, 64–66; and Paret, *The Cognitive Challenge of War*, 132–133.
24. It was not until the 1976 edition by Howard and Paret that the original German-language text was used as the basis for an English edition; earlier translators relied on already corrupted versions of the work. Bassford, *Clausewitz in English*, 57–58.
25. Muth, *Command Culture*, 20.
26. Strachan, *Clausewitz's On War*, 102.
27. Many consider the 1943 Otto Jolles translation to be the most accurate of all English-language editions. However, it has been criticized for its literalness, which can make it difficult to read. Jolles, an academic with no military experience—he undertook the translation during World War II in part to avoid the draft—may also have missed some contextual nuances. Despite being largely forgotten today, the Jolles version receives high praise from linguistic experts for its fidelity to the original text. It presents an intriguing "what if" for Clausewitz scholars, as it offered a superior translation of *On War* more than three decades before Howard and Paret. Bassford, *Clausewitz in English*, 172, 181, 183–185, and 193.
28. Sumida, *Decoding Clausewitz*, 46–47; and Clausewitz, *On War* (Graham/Rapoport).
29. Fleming, *Clausewitz's Timeless Trinity*, 61.
30. Clausewitz, *On War* (Graham/Rapoport), 21; and Bassford, *Clausewitz in English*, 58–59 and 198.
31. Aron, *Clausewitz: Philosopher of War*, 102, 104, 223, 315–316, 328, 340, and 408.

32. Like Clausewitz, in 1813, Jomini also decided to join the Russian army to oppose Napoleon. Bassford, *Clausewitz in English*, 16. See also, Stoker, *Clausewitz: His Life and Work*, 137.

33. Howard, *Clausewitz*, 59. This discrepancy was exacerbated further still by the early death of Marie von Clausewitz in 1836 as she was the greatest promoter and defender of her husband's legacy. Bellinger, *Marie von Clausewitz*, 12.

34. Sumida, *Decoding Clausewitz*, 12–13.

35. Bassford, *Clausewitz in English*, 17–18.

36. Jomini, *The Art of War* (Mendell and Craighill); and Bassford, *Clausewitz in English*, 36.

37. Reardon, *With a Sword in One Hand and Jomini in the Other*.

38. Paret, *The Cognitive Challenge of War*, 129.

39. Jomini, *The Art of War*, 17

40. Waldman, *War, Clausewitz, and the Trinity*, 105.

41. Jomini, *The Art of War*, 14.

42. Bassford, *Clausewitz in English*, 111.

43. Reardon, *With a Sword in One Hand and Jomini in the Other*; and Summers, "Introduction" in *War, Politics, and Power*, xi.

44. Clausewitz, *On War* (Howard and Paret), II.2, 134. See also Waldman, *War, Clausewitz, and the Trinity*, 21,

45. Italics in original; Clausewitz, *On War* (Howard and Paret), II.2, 135. See also Howard, *Clausewitz*, 41.

46. Clausewitz, *On War* (Howard and Paret), II.2, 141. See also Paret, *The Cognitive Challenge of War*, 127–128.

47. Clausewitz, *On War* (Howard and Paret), VI.6, 374.

48. Winston Churchill, "Finest Hour Speech," June 18, 1940, https://winstonchurchill.org/resources/speeches/1940-the-finest-hour/their-finest-hour/.

49. On the urgency of beating the Nazis in the race to develop atomic weapons, see Rhodes, *The Making of the Atomic Bomb*, esp. 3–4, 227, 264–271, 273, 284–288, 297, 311, 319–323, 347, 384–385, and 432.

50. Stoker, *Clausewitz: His Life and Work*, 283.

51. Howard, *Clausewitz*, 70–71.

52. Brodie, *The Absolute Weapon*.

53. Echevarria II, *Clausewitz and Contemporary War*, 85–86. More recently, scholars, such as Stephen Cimbala, have refined and expanded these theories using the improved Paret and Howard translation. See Cimbala, *Clausewitz and Escalation*.

54. While Clausewitzian thought is sprinkled throughout Brodie's works, the most refined example is Brodie, *War and Politics*, 1–2, 8–11, 13, 15, 37–38, 47, 262, 436, 438, 440–447, 449, 451–453, 475, and 494.

55. See, for example, Kahn, *Thinking About the Unthinkable*; Kissinger, *Nuclear Weapons and Foreign Policy*; Osgood, *Limited War*; and Schelling, *Arms and Influence*.

56. Bassford, *Clausewitz in English*, 114, 97, 200–204, Strachan, *Clausewitz's On War*, 147 and 152, and Sumida, *Decoding Clausewitz*, 187.

57. An extremely short abridgement of his work was issued by the United States Military Academy at West Point as early as 1943, in the volume "Jomini, Clausewitz, and Schlieffen" but there is no evidence that it was an official part of the curriculum. Muth, *Command Culture*, n39 231.

58. Stoker, *Clausewitz: His Life and Work*, 284 and Summers, "Introduction," ix–xv.

59. Kitfield, *Prodigal Soldiers*.

60. On the adoption of Thucydides by American Professional Military institutions, see Stansfield Turner, "Address to Chicago Council Navy League of the United States," March 9, 1973, https://www.cia.gov/readingroom/docs/CIA-RDP80B01554R003500280001-7.pdf; and Novo and Parker, *Restoring Thucydides*, 37–38 and 73.

61. Strachan, *Clausewitz's On War*, 1–2.

62. It is unclear which translation of *On War* Powell actually read. Since this incident occurred in 1975, it could not have been the now-standard 1976 translation by Howard and Paret. Powell and Persico, *My American Journey*, 207–208; and Strachan, *Clausewitz's On War*, 1–2.

63. Heuser, *Reading Clausewitz*, 176.

64. Bassford, *Clausewitz in English*, 3. For an argument that Weinberger and Powell may have misunderstood Clausewitz's conception of the trinity and perpetuated an overly state-centric view, see Fleming, *Clausewitz's Timeless Trinity*, 51–52.

65. United States Marine Corps, *FMFM 1: Warfighting*, https://www.marines.mil/Portals/1/Publications/MCDP 1 Warfighting.pdf. See also Bassford, *Clausewitz in English*, 204–205.

66. For an example of Clausewitz's work being used to analyze failures in Vietnam immediately prior to the 1976 publication of Howard and Paret's translation becoming widely available, see Staudenmaier, *Vietnam*.

67. Bassford, *Clausewitz in English*, 207–208.

68. Strachan, "Michael Howard and Clausewitz," 143–160; and Sumida, "A Concordance," 271–331. For links to various indexes for *On War*, see "Indexing *On War*" at https://clausewitzstudies.org/.

69. Strachan, *Clausewitz's On War*, 1.

70. Bassford, *Clausewitz in English*, 4–5.

71. Summers, *On Strategy*, 1.

72. Summers, esp. 1, 21–31, and 33–80.

73. Summers, 43–44.

74. Summers makes this point even more explicitly in Summers, "Introduction," x.

75. Summers focused his analysis on the second formulation of the trinity (the people, the military, and the government), but largely ignored the original formulation of the trinity (primordial violence, hatred, and enmity). Herberg-Rothe, *Clausewitz's Puzzle*, 6.

76. Despite its considerable flaws, Summers's book sold well, and it was updated and released in paperback, even spawning a similar treatment of the 1991 Gulf War; see Summers, *On Strategy II*. See also Heuser, *Reading Clausewitz*, 52 and 165–175; Strachan, *Clausewitz's On War*, 2 and 178–179; and Waldman, *War, Clausewitz, and the Trinity*, 166.

77. See, for example, Clodfelter, *The Limits of Airpower*, xi–xii, 1, 203, and 206; Cohen and Gooch, *Military Misfortunes*, 2, 24, 44–45, 190, and 195; Griffith, *Battle Tactics of the Civil War*, 29, 45, 63, and 198; and Roland, *An American Iliad*, 40, 44–45, 68, 181, and 262. On this point, see also Bassford, *Clausewitz in English*, 209.

CHAPTER 7

THE CORE CONCEPTS OF *ON WAR*: CLAUSEWITZ'S FOUR CRITICAL INSIGHTS

Nearly all military officers and armchair strategists know (or think they know) Clausewitz's four most famous concepts: friction, war as an extension of politics, the tendency of violence to escalate toward extremes, and the trinity. Yet despite their widespread use, these four key concepts are often misunderstood. Like war itself, they appear simple but are in fact deeply complex.[1] This chapter explains these four central concepts and how they interconnect within Clausewitz's theory of war.

Friction, Complexity, and Uncertainty

The concept of friction is one of Clausewitz's most important and enduring contributions to the study of war. While this theory is complex, it is also highly intuitive. At its core, it captures the idea that unexpected problems inevitably arise. Anyone who has experienced war or participated in military planning can attest to how often even the best-laid plans go awry. Despite the common-sense understanding that war is inherently imperfect, many theories overlook this fundamental insight. Clausewitz believed this was a critical oversight and built his theory of war around a tragic view of human imperfection.[2]

Friction appears to have been the earliest of Clausewitz's major theoretical insights, first emerging during the opening stages of the 1806 campaign. In a letter to Marie dated September 29 that same year, he lamented the difficulties that General Scharnhorst encountered simply in moving his troops:

> How much must the effectiveness of a gifted man be reduced when he is constantly confronted by the obstacles of convenience and tradition, when he is paralyzed by constant friction with the opinions of others.[3]

While Clausewitz idolized Scharnhorst, he could see that even Scharnhorst's energy and genius were diminished by complex forces beyond his control. From this moment forward, Clausewitz encountered countless opportunities to witness friction firsthand, including the Prussian campaign of 1806, Napoleon's invasion of Russia in 1812, the "damn near run thing" that was the Battle of Waterloo in 1815, and the repeated delays and impediments to his own plans for the Prussian War College. Clausewitz channeled these experiences and frustrations into his philosophy of war. He understood that chance, uncertainty, and complexity were inherent in all human endeavors, yet military theorists often neglected these factors in their analyses.[4] As a result, they perpetuated unrealistic theories that looked sound on paper but were dangerously divorced from reality.[5]

In describing the importance of friction to his understanding of war, Clausewitz emphasized that it could not be fully conveyed to those who have not experienced it firsthand:

> Friction is the only concept that more or less corresponds to the factors that distinguish real war from war on paper...the dangers inseparable from war and the physical exertions war demands

can aggravate the problem to such an extent that they must be ranked among its principal causes.[6]

Unlike Jomini and other theorists of his time, Clausewitz neither promised victory nor offered a simple checklist for success. From its very inception, his theory was constructed to grapple with complexity rather than reduce it.

For Clausewitz, friction is a relatively simple concept with profound implications. All human actions fall short of perfection; events are much more difficult and complicated than we would like to believe and, no matter how hard we try, we can never fully eliminate the effects of friction:

> Everything in war is very simple, but the simplest thing is difficult. The difficulties accumulate and end by producing a kind of friction that is inconceivable unless one has experienced war.[7]

Clausewitz continues with an extended analogy of a traveler at night:

> Imagine a traveler who late in the day decides to cover two more stages before nightfall. Only four or five hours more, on a paved highway with relays of horses: it should be an easy trip. But at the next station he finds no fresh horses, or only poor ones; the country grows hilly, the road bad, the night falls, and finally after many difficulties he is only too glad to reach a resting place with any kind of primitive accommodation. It is much the same in war. Countless minor incidents—the kind you can never really foresee —combine to lower the general level of performance, so that one always falls far short of the intended goal.[8]

For Clausewitz, war always existed within the realm of chance and uncertainty, and it would be foolish to think otherwise.[9]

He grounded his understanding of friction in a "thick" view of human nature. According to this view, human beings are driven by a wide range of emotions, including fear, honor, interest, love, patriotism, greed, and hatred. Foremost among these was fear.[10] Clausewitz illustrated the power of fear in a vivid passage drawn from his own battlefield experiences:[11]

Let us accompany a novice to the battlefield. As we approach the rumble of guns grows louder and alternates with the whir of cannonballs, which begin to attract his attention. Shots begin to strike close around us. We hurry up the slope where the commanding general is stationed with his large staff. Here cannonballs and bursting shells are frequent, and life begins to seem more serious than the young man imagined. Suddenly someone you know is wounded; then a shell falls among the staff. You notice that some of the officers act a little oddly; you yourself are not as steady and collected as you were: even the bravest can become easily distracted. Now we enter the battle raging before us, still almost like a spectacle, and join the nearest divisional commander. Shot is falling like hail, and the thunder of our own guns adds to the din. Forward to the brigadier, a soldier of acknowledged bravery, but he is careful to take cover behind a rise, a house or a clump of trees. A noise is heard that is a certain indication of increasing danger—the rattling of grapeshot on roofs and on the ground. Cannonballs tear past, whizzing in all directions, and musketballs begin to whistle around us. A little further we reach the firing line, where the infantry endures the hammering for hours with incredible steadfastness. The air is filled with hissing bullets that sound like sharp cracks if they pass close to one's head. For a final shock, the sight of men being killed and mutilated moves our pounding hearts to awe and pity...Danger is part of the friction of war. Without an accurate conception of danger we cannot understand war. That is why I have dealt with it here.[12]

For Clausewitz, fear by its very nature was an emotion that could never be completely overcome and would ensure that war was never a perfectly rational enterprise.

Based on this understanding of human nature, Clausewitz concluded that imperfection is an inescapable part of the human condition. It flows logically that, as a human enterprise, war can never be perfected. Rather than attempting to eliminate friction or achieve perfection, the best a strategist can ever hope for is to anticipate and mitigate imperfection—to be less burdened by friction than their opponents.[13] To this end,

Clausewitz begins his section on planning with a powerful reminder that real war is inherently complex:

> On the one hand, military operations appear extremely simple. The greatest generals discuss them in the plainest and most forthright language; and to hear them tell how they control and manage that enormous, complex apparatus one would think the only thing that mattered was the speaker, and that the whole monstrosity of war came down, in fact, to a contest between individuals, a sort of duel. A few uncomplicated thoughts seem to account for their decisions—either that, or the explanation lies in various emotional states; and one is left with the impression that great commanders manage matters in an easy, confident and, one would think, offhand sort of way. At the same time we can see how many factors are involved and have to be weighed against each other; the vast, the almost infinite distance there can be between a cause and its effect, and the countless ways in which these elements can be combined.[14]

While this may seem like an obvious point for those who have experienced conflict, it is frequently forgotten by those responsible for planning strategy and making crucial decisions regarding war and peace.[15] Although much of *On War* discusses the best tactics—such as optimizing geographical positioning, protecting the flanks, employing firepower, and bolstering morale—these could only improve the odds; they could never guarantee victory. Clausewitz had witnessed firsthand the disastrous effects of overconfidence on the battlefield, and he emphasized that because of friction, the outcome of a battle could never be known with certainty in advance.

For some, Clausewitz's view of human imperfection might seem profoundly disappointing, even defeatist. Unlike Jomini who promised victory if his methods were followed, Clausewitz's concept of friction led him to adopt the language of probability rather than certainty. One could calculate and plan, but an element of chance would always persist in any human endeavor. To illustrate this probabilistic dynamic, Clausewitz employed the analogy of gambling:

> absolute, so-called mathematical, factors never find a firm basis in military calculations. From the very start there is an interplay of possibilities, probabilities, good luck and bad that weaves its way throughout the length and breadth of the tapestry. In the whole range of human activities, war most closely resembles a game of cards.[16]

This does not mean knowledge is impossible or that planning is irrelevant. Rather, the most one can hope for are probabilistic assessments of the likelihood of success or failure in any given undertaking.[17]

This probabilistic view of war is uncomfortable, particularly for leaders who are uncertain or risk-averse. It offers no guaranteed solutions or predictions, only judgments about the most likely outcomes. One can work diligently to improve the odds, but certainty will always remain elusive. As Clausewitz observed, "In war ...all action is aimed at probable rather than at certain success. The degree of certainty that is lacking must in every case be left to fate, chance, or whatever you like to call it."[18]

This inability to achieve certainty led Clausewitz to adopt a view of fortune strikingly similar to that of Thucydides and Machiavelli. According to this classical view, fortune is an omnipresent force in human affairs. Fortune does not choose sides, and it is ready to be seized by a bold and decisive leader.[19] This does not imply that a leader should act recklessly; rather, Clausewitz recognized that leaders would never possess complete information or ideal conditions, and that bold action was often necessary for success. Waiting for a perfect solution is a recipe for failure as the perfect is often the enemy of the good:

> we should not habitually prefer the course that involves the least certainty. That would be an enormous mistake...There are times when the utmost daring is the height of wisdom.[20]

In Clausewitz's classical conception of leadership, the ability to assess a situation rapidly and act decisively cannot eliminate friction but can cut through the mental fog of indecision and provide a critical advantage amid the chaos and confusion of battle.[21]

When discussing Clausewitz's understanding of friction, it is important to be precise about what he actually wrote rather than relying on popular interpretations. One of the most enduring misconceptions is that Clausewitz coined the phrase, "fog of war" as a catch-all phrase for complexity and chaos. Despite this widespread belief, Clausewitz never used the term "fog of war" in the way it is commonly understood today. Instead, he referred to *fog* in two contexts: as a literal weather condition and as a metaphor for mental confusion. In both cases, however, his usage differs from the modern concept. For Clausewitz, fog was not a foundational concept in itself, but rather a descriptive element illustrating how friction can hinder military operations.

In *On War*, Clausewitz described weather phenomena such as rain, snow, or actual fog as examples of chance events that could introduce friction into military operations.[22] He observed, "Fog can prevent the enemy from being seen in time, a gun from firing when it should, a report from reaching the commanding officer." While Clausewitz generally downplayed the significance of weather in shaping the outcome of military affairs, he made a notable exception for fog, stating, "It is rarer still for weather to be a decisive factor. As a rule only fog makes any difference."[23]

The second use of the term *fog* refers to the difficulty of decision-making under the stress and uncertainty of battle. For Clausewitz, this uncertainty demanded a strength of character and moral resilience that few possess.[24] Successful leaders must confront and grow comfortable with the reality that:

> War is the realm of uncertainty; three quarters of the factors on which action in war is based are wrapped in a fog of greater or

lesser uncertainty. A sensitive and discriminating judgment is called for; a skilled intelligence to scent out the truth.[25]

This seeming inconsistency and illogic present a psychological challenge for a general to master because "in war everything is uncertain, and calculations have to be made with variable quantities...[and] a continuous interaction of opposites."[26] This psychological burden is compounded further as "all action takes place, so to speak, in a kind of twilight, which like fog or moonlight, often tends to make things seem grotesque and larger than they really are."[27]

In both of these uses—weather and uncertainty—Clausewitz employs fog as an example of friction and complexity that must be understood and overcome by a successful general. This is part of Clausewitz's broader approach to friction, uncertainty, and military genius, but it should not be understood as a single, stand-alone theory. Unfortunately, the phrase "fog of war" is often one of the few lines attributed to Clausewitz that people remember, along with the frequently quoted maxim that "war is the continuation of politics by other means," and is therefore one of the most misunderstood.

These fundamental misunderstandings and oversimplifications of Clausewitz's theories have led to some potentially dangerous, albeit often well-intentioned, conclusions. In recent decades, some theorists have argued that advances in smart weapons, communications technologies, and battlefield sensors can overcome friction.[28] They claim that improved information and precision-strike capabilities can make war cleaner, faster, and more controllable. While such claims may be appealing, Clausewitz would have viewed them with deep skepticism. He insisted that war is inherently violent and uncertain, and he warned that more information often compounds uncertainty rather than resolving it. Addressing the dangers of information overload and analysis paralysis, Clausewitz observed:

we know more, but this makes us more, not less, uncertain...The latest reports do not arrive all at once: they merely trickle in. They continually impinge on our decisions, and our mind must be permanently armed, so to speak, to deal with them.[29]

In short, advances in information technology and precision targeting do not fundamentally alter the nature of war, and it would be foolish to believe otherwise.

Paradoxically, more information and more accurate weapons might exacerbate the problems that they were intended to solve by introducing additional layers of complexity.[30] War is dynamic and friction is omnipresent, which means the enemy will adapt, machines will fail, and new problems will inevitably emerge. In short, war will continue as a perpetual series of moves and countermoves, just as Clausewitz anticipated. In such a world, friction will always find ways to limit and degrade human actions, and perfection will remain unobtainable. Strategists must therefore approach war with humility and always account for friction in their understanding of its nature.

War and Politics

Unfortunately, the most famous quote attributed to Clausewitz is also one of its most misunderstood: "War is the continuation of politics by other means."[31] For many of Clausewitz's critics, including Anatol Rapoport, Basil Liddell Hart, John Keegan, and Barbara Tuchman, this phrase exemplifies a brutalist form of Prussian militarism. According to this interpretation, when peaceful politics fail to yield the desired results, war becomes the logical and necessary next step.[32] Clausewitz, however. would likely have been dismayed by such a reading of his work.[33]

Fortunately, this common misunderstanding is relatively simple to correct. To grasp Clausewitz's intended meaning, it is essential to read the full passage rather than the often-cited, oversimplified version:

> We see, therefore, that war is not merely an act of policy but a true political instrument, a continuation of political intercourse, carried on with other means. What remains peculiar to war is simply the peculiar nature of its means. War in general, and the commander in any specific instance, is entitled to require that the trend of designs of policy shall not be inconsistent with these means...The political object is the goal, war is the means of reaching it, and the means can never be considered in isolation from their purpose.[34]

In the simplest form, Clausewitz's point is that nations use various instruments of power—diplomatic, informational, military, and economic —to achieve their political objectives.[35]

While Clausewitz focused his work on the military instrument, he never claimed that it was the only tool or even the best tool. Rather, he recognized the brutality of war, and repeatedly advocated restraint:

> War is not pastime; it is no mere joy in daring and winning, no place for irresponsible enthusiasts. It is a serious means to a serious end, and all its colourful resemblance to a game of chance, all the vicissitudes of passion, courage, imagination, and enthusiasm it includes are merely its special characteristics.[36]

With this warning, Clausewitz made clear that war was not a sport or a source of excitement, but a deadly serious undertaking.[37]

According to Clausewitz, successful strategists must understand the purpose of the war, focus their energies on achieving the desired ends, and avoid distraction:

> If we keep in mind that war springs from some political purpose, it is natural that the prime cause of existence will remain the supreme consideration in conducting it...Policy, then will permeate all military operations, and, in so far as their violent nature will admit, it will have a continuous influence on them.[38]

While this is difficult to achieve in practice, the theory itself is remarkably simple. In fact, this understanding of the need to balance ends and means is so deeply embedded in the vernacular of statecraft that it is almost banal. Of course, it is essential to understand the intended goal and to ensure that military force is used in a way that aligns with that goal.

Clausewitz was not so naïve as to believe that war could be confined to a purely rational exercise of balancing ends and means. On the contrary, he understood that war was a profoundly human activity —chaotic and difficult to control—and he built his theory to account for this dynamic. Consistent with his dialectical approach, Clausewitz presented the rationalist approach to war as an ideal type of how strategists *should* act: "No one starts a war—or rather, no one in his senses ought to do so —without first being clear in his mind what he intends to achieve by the war and how he intends to conduct it."[39]

In practice, Clausewitz acknowledged that properly balancing ends and means during war is exceptionally difficult. Clausewitz's nemesis, Napoleon, exemplified this dilemma: though a brilliant general who achieved a series of dramatic victories, Napoleon frequently undermined his own political objectives by pursuing ever more ambitious war aims. These escalating goals provoked further political unrest, additional conflicts, and increasingly brutal levels of violence. In short, Napoleon failed because he adopted progressively unreasonable policies. Drawing on this understanding of history and human nature, Clausewitz expanded his theory to explain how war could take on a life of its own and cease to be a purely rational enterprise.

Violence Tends Toward Extremes
The third of Clausewitz's critical insights is his observation that war tends toward extreme levels of violence. Unfortunately, much like the related observation that war is a political act, this theoretical insight has frequently been misquoted and misappropriated. Indeed, *On War* is replete with statements that, when taken out of context, can appear to

advocate for unlimited violence in the pursuit of policy objectives. For example, Clausewitz described the new wars unleashed by the French Revolution as "untrammelled by any conventional restraints...broken loose in all its elemental fury."[40] He asserted that "the maximum use of force is in no way incompatible with the simultaneous use of the intellect."[41] He also scorned "kind-hearted people" who "think there was some ingenious way to disarm or defeat an enemy without too much bloodshed."[42] Perhaps most famously, he theorized:

> The thesis, then, must be repeated: war is an act of force, and there is no logical limit to the application of that force. Each side, therefore, compels its opponent to follow suit; a reciprocal action is started which must lead, in theory to extremes.[43]

On the surface, these quotes might appear to be the ravings of a madman who fetishized war and reveled in violence. Indeed, the German nationalist and proto-fascist Friedrich von Berhardi hijacked Clausewitz's theories to argue that war should be taken to the limits of utmost violence and that Germany should preemptively destroy other nations before they could strike first.[44] However, this interpretation could not be further from the truth.

To understand Clausewitz's claim, it is important to remember that he employed a dialectical approach and held a tragic view of human nature. In Book I, Chapter 1 of *On War*, he begins with a thesis about the theoretical nature of war tending toward extremes. Logically, war is about winning through the application of superior force, disarming the enemy, and responding dynamically to an adversary's moves. Thus, victory becomes a matter of inflicting greater violence than one's opponent, leading to a spiral of escalation until one side is destroyed. Clausewitz concludes that these factors drive war toward ever greater levels of violence and force.[45]

After presenting a theoretical model in which war tends toward extremes of violence, Clausewitz immediately complicates this view by demonstrating its practical limitations. His antithesis contends that war in reality is rarely, if ever, truly unlimited or extreme for three primary reasons. First, military and political leaders always operate under constraints that limit the use of force. Second, because a nation must retain some strength in reserve, it can never commit its full capabilities at once. Third, because total victory was unattainable and defeat rarely permanent, rational leaders are more likely to pursue a negotiated peace that is acceptable to both sides than to attempt the complete destruction of the enemy.[46]

Rather than offering a neat synthesis, Clausewitz concludes that although war tends toward extremes, it is never truly absolute: "Warfare thus eludes the strict theoretical requirement that extremes of force be applied." Since this theoretical maximum is never reached, "it becomes a matter of judgement what degree of effort should be made."[47] Once judgement enters the equation, war becomes an exercise in calculation and pursuit of political interests. This political purpose, Clausewitz explains, will inevitably reassert itself: "as this determination [for maximum violence] wanes, the political aim will reassert itself."[48]

The real world, therefore, aligns with neither extreme but perpetually fluctuates between these two poles:

> War is a pulsation of violence, variable in strength and therefore variable in the speed with which it explodes and discharges its energy. War...always lasts long enough for influence to be exerted on the goal and for its own course to be changed in one way or another...If we keep in mind that war springs from some political purpose, it is natural that the prime cause of its existence will remain the supreme consideration in conducting it. That, however, does not imply that the political aim is a tyrant. It must adapt itself to its chosen means, a process which can radically change it; yet the political aim remains the first consideration. Policy, then, will permeate all military operations, and, in so far as their violent nature will admit, it will have a continuous influence on them.[49]

For those seeking a definitive answer about the nature of war, this conclusion may seem deeply unsatisfying as it offers neither a clear prediction nor a simple prescription.

The challenge for strategists is to recognize that while there is pressure to escalate violence to ever-increasing extremes, natural limits constrain the use of force, and reason and sound judgement must ultimately guide policy. On the surface, this may appear self-evident. However, in the aftermath of the extreme violence of the Napoleonic Wars, Clausewitz's warning was particularly relevant.[50] He had witnessed firsthand how Napoleon had repeatedly expanded his war aims to increasingly ambitious and destructive levels, a strategy that ultimately led to his downfall.[51] As Clausewitz predicted, this spiral of violence is difficult to control and becomes self-defeating because it ultimately undermines the very political purpose for which the war was undertaken.

To understand this dynamic, Clausewitz returns to his theories of friction and human nature. He believed that wars are inherently complex and imperfect, fought by flawed humans who are not always rational. Rather than attempting to eliminate friction or alter human nature, he acknowledged the tendency toward extremes but advocated wisdom and restraint. War, he warned. should not be an "act of senseless passion."[52] Instead, wise strategists must try to master their animalistic instincts and conduct war in a manner:

> controlled by its political object, the value of the object must determine the sacrifices to be made for it in *magnitude* and also in *duration*. Once the expenditure of effort exceeds the value of the political object, the object must be renounced and peace must follow (italics in the original).[53]

When viewed in this manner, Clausewitz's theories of war tending toward extremes offers an elegant way for understanding the destruction wrought by the French Revolution and the Napoleonic Wars. These conflicts, fought in pursuit of extreme political aims, escalated to correspondingly extreme levels of violence. Clausewitz's argument is not merely an explanation for their destructiveness; it is also a caution to policymakers to consider the consequences of their actions and exercise rationality and restraint.

Clausewitz returns to the theme of managing the escalation of violence as a political challenge throughout the remainder of his work, most notably in Book 8, Chapter 3. Under the heading, "Scale of the Military Objective and the Effort to be Made," he reminds the reader that while it may be tempting to pursue maximum levels of violence in the false hope that this will achieve the desired ends, such cycles of escalation are ultimately self-defeating:

> Since in war too small an effort can result not just in failure but in positive harm, each side is driven to outdo the other, which sets up an interaction. Such an interaction could lead to a maximum effort if a maximum could be defined. But in that case all proportion between action and political demands would be lost; means would cease to be commensurate with ends, and in most cases a policy of maximum effort would fail because of the domestic problems it would raise.[54]

Clausewitz again reminds the reader to avoid becoming a prisoner of passion and instead act rationally and "adopt a middle course...on the principle of using no greater force and setting himself no greater aim than would be sufficient for the achievement of his political purpose."[55]

While Clausewitz's discussion of war tending toward extremes can at times seem repetitive and awkward, it was central to his theoretical and moral framework. From a moral standpoint, Clausewitz was appalled by the unrestrained violence he witnessed during the Napoleonic Wars, which he saw as unnecessary and counterproductive. His fear of extreme violence and repeated pleas for restraint were deeply personal and sincere. Logically, his theoretical claim that war never reaches its absolute extreme serves as a central intellectual thread connecting his ideas. By modeling war in this way, Clausewitz highlights the omnipresence of friction and imperfection, explains the difficulty of balancing ends and means, reinforces his view of war as a political act, and lays the foundation for his most famous theoretical construct: the trinity.

Trinity/Trinities

The fourth of Clausewitz's well-known but frequently misunderstood theories is his concept of the "paradoxical trinity."[56] According to this theory, war is defined by the interplay of violent emotion, chance, and rational calculation. These three forces exert constant and competing pressures; while one may dominate at a particular moment, the system remains in flux, and no single element ever holds complete sway. This model accounts for both continuity and change: the constituent elements remain constant, but their interaction is always evolving. By describing war in this manner, Clausewitz lends intellectual coherence to his broader theory, synthesizing his thesis that war was a rational political act with his antithesis that war tends toward extremes.[57] War, therefore, is never entirely political or fully rational, nor is it taken to its absolute extreme.

Clausewitz had been fascinated by these apparent contradictions since his early days as a junior officer during the Napoleonic Wars. He observed how Napoleon fundamentally transformed European warfare by mobilizing the energies of the people and reinforcing them with the growing power of the nation-state.[58] Rather than the limited conflicts waged by the professional armies of Frederick the Great, Napoleon's style of warfare was orders of magnitude larger and more destructive. The defeats at Jena and Auerstädt in 1806 fundamentally shattered the old paradigm of limited wars and forced the Prussian elite to reconsider their understanding of armed conflict. As early as 1807, just one year later, Clausewitz had begun to develop the core insight that would eventually lead him to his theory of the trinity. His writings from this period explore the interplay between passion and luck, and the effects these two elements had on the conduct of war. Although he would later revise "luck" to "chance" and add "rationality" as a third component, it is evident that he had already achieved a theoretical breakthrough.[59] War was continually shapedby the persistent tension among opposing forces.

The understanding that war is always in a state of flux was central to Clausewitz's concept of the trinity. In *On War*, Clausewitz introduces the trinity as a pull between forces, not unlike the thrust and parry of duelists, each seeking advantage over the other. He then describes this dynamic in greater detail, defining the trinity as follows:

> War is more than a true chameleon that slightly adapts its char-
> acteristics to the given case. As a total phenomenon its dominant
> tendencies always make war a paradoxical trinity—composed of
> primordial violence, hatred, and enmity, which are to be regarded
> as a blind natural force; of the play of chance and probability
> within which the creative spirit is free to roam; and of its element
> of subordination, as an instrument of policy, which makes it
> subject to reason alone.[60]

This initial conception of the trinity reflects Clausewitz's views that war is an intensely human act, driven by intensely human emotions. The recent Napoleonic Wars were therefore an extreme example of this dynamic because they tapped into the pent-up energies of the people and, for a time, exceeded what many believed to be probable or rational.

While Clausewitz believed that these three forces—passion, probability, and reason—offered the best framework for understanding the shifting face of war, he also recognized that this model would be unsatisfying without a concrete link between these elementsand the principal actors involved in war. To ground these abstract forces in institutional reality, Clausewitz introduced a corresponding trinity:

> The first of these aspects mainly concerns the people; the second
> the commander of the army; the third the government. The
> passions that are to be kindled in war must already be inherent
> in the people; the scope which the play of courage and talent
> will enjoy in the realm of probability and chance depends on
> the particular character of the commander and the army; but the
> political aims are the business of government alone.[61]

In many ways, this second formulations of the trinity appears more intuitive and easier to apply than the emotionally charged and conceptually complex first version.[62] As a result, many readers mistakenly treat Clausewitz's two formulations as interchangeable, overlooking his careful use of the phrase "mainly concerns." This oversimplification is understandable —readers often seek straightforward explanations—but it has been unfortunately reinforced by influential critics of Clausewitz, including Harry Summers, Martin van Creveld, and Mary Kaldor.[63]

A closer reading reveals that Clausewitz did not regard these two trinities as equivalent.[64] Rather, the abstract concepts he initially introduced—passion, chance, and reason—constitute the true trinity, while the second formulation represents an attempt to translate these emotional forces into their rough institutional analogues. By using the word "mainly," Clausewitz signaled that these two formulations are related but not interchangeable. Equating them misses the essential point: Clausewitz was making a complex emotional argument about the nature of war, not a strictly rational, state-centered one. Adopting the formulation of people, military, and government can lead to the mistaken belief that successful policy can be achieved merely by coordinating these three instruments of power. While Clausewitz advocated waging war as rationally as possible, he fundamentally rejected the idea that improved orchestration or greater care could overcome the irrational pull of human emotion.[65] Indeed, Clausewitz's core insight is that the nature of war is neither fixed nor fully rational.

For the policymaker, the challenge is far more complex than simply making a decision that maximizes expected utility and balances the three elements of the trinity. Rather, the task is uncertain, complex, and dynamic as the policymaker must pursue objectives while navigating the constant pull of powerful emotions.[66] This accurate understanding of the trinity is harder to apply precisely because it is grounded in intangible, emotion-driven forces rather than rational constructs. Although this emotion-based interpretation of the trinity is more difficult to comprehend and manage, it more accurately reflects Clausewitz's original intent and aligns more closely to his broader philosophical approach.

Clausewitz's theory presents the frustrating reality that all three elements are present simultaneously, and while one may be more important in a given moment or situation, none of the three elements ever truly dominates the others.[67] Clausewitz warned:

> These three tendencies are like three different codes of law, deep-rooted in their subject and yet variable in their relationship to one another. A theory that ignores any one of them or seeks to fix an arbitrary relationship between them would conflict with the reality to such an extent that for this reason alone it would be totally useless.[68]

Rather than one element dominating the others, there is a continual pull among these conflicting poles. Clausewitz acknowledged that this perpetual tension and flux made theory building and policy implementation considerably more difficult stating, "Our task therefore is to develop a theory that maintains a balance between these three tendencies, like an object suspended between three magnets."[69] This perpetual and unresolved tension ensures that no single part of the trinity ever fully dominates the others and that the strategic landscape remains in constant flux.[70]

Due to the constantly shifting strategic landscape, the trinity should not be treated as a three-part checklist for policymakers.[71] Rather, strategists should recognize these tensions, resist their siren song, and chart a wise and prudent course between competing extremes.[72] This is an extremely difficult and often frustrating task, and there will always be a temptation to adopt a more reductive and linear version of Clausewitz's complex theory.

Ultimately, to grasp these four concepts, readers must accept that they are inherently complex and seemingly contradictory. War is defined by friction and is probabilistic, though never certain. It is an act of force with political purpose, but this political purpose may shift over time. War tends toward extremes but is never truly extreme. The trinity is an imperfect model for representing the complex and perpetual interplay of human emotions. These are difficult ideas, and it may be tempting to dismiss them as impractical. Yet they are not as opaque as they might initially appear and can offer a lifetime of insight and inspiration for those committed to deeper understanding.

NOTES

1. Echevarria, *Clausewitz and Contemporary War*, 1.
2. Lebow, *The Tragic Vision of Politics*, esp. 168–215.
3. Bellinger, *Marie von Clausewitz*, 70–71; Echevarria, *Clausewitz and Contemporary War*, 103; Paret, *Clausewitz and the State*, 124 and 197; and White, *The Enlightened Soldier*, 127–128.
4. Lebow, *The Tragic Vision of Politics*, 195.
5. Waldman, *War, Clausewitz, and the Trinity*, 103.
6. Clausewitz, *On War* (Howard and Paret), *On War*, I, 7, 119.
7. Clausewitz.
8. Clausewitz.
9. Forgetting this fact fosters a false sense of certainty about the prospects of victory, thereby increasing the likelihood of war. Blainey, *The Causes of War*, 194.
10. For a perspective that views emotions such as fear and anger as natural and potentially constructive for warriors, see Sherman, *Stoic Warriors*, esp. 67 and 71; and Waldman, *War, Clausewitz, and the Trinity*, 160.
11. Engberg-Pedersen, *Empire of Chance*.
12. Clausewitz, *On War* (Howard and Paret), I.4, 113–114.
13. For attempts to shift friction to an opponent as a means of gaining military advantage, see Boyd, *The Essence of Winning and Losing*; and Watts, *Clausewitzian Friction and Future War*.
14. Clausewitz, *On War* (Howard and Paret), VIII.1, 577.
15. Dunnigan, *How to Make War*, 329–341. See also Blainey, *The Causes of War*, and Kaplan, *Daydream Believers*. For an argument that Clausewitz's concept of friction is unremarkable, see van Creveld, "The Transformation of War Revisited," 3–15.
16. Clausewitz, *On War* (Howard and Paret), I.1, 86.
17. In certain respects, Clausewitz's thinking here anticipates the work of twentieth-century business theorists who distinguished between quantifiable risk and unquantifiable uncertainty. For the classic treatment of this distinction, see Knight, *Risk, Uncertainty, and Profit*.
18. Clausewitz, *On War* (Howard and Paret), II.5, 167.
19. This theme runs throughout Machiavelli's work. For one of its clearest expressions, see "One Often Obtains with Impetuosity and Audacity What One Would Never Have Obtained through Ordinary Modes" in

Machiavelli, *Discourses on Livy* (Mansfield and Narcov), 304; and Machiavelli, *The Prince* (Mansfield), 98–100. See also Waldman, *War, Clausewitz, and the Trinity*, 129–131.

20. Clausewitz, *On War* (Howard and Paret), II.5, 167.
21. Beyerchen, "Clausewitz, Nonlinearity, and the Unpredictability of War," 59–90; Bousquet, *The Scientific Way of Warfare*, 10, 33, 81, 85–90, 156, 197, 236, and 241; Herberg-Rothe, *Clausewitz's Puzzle*, 81–82; Jervis, *System Effects*, 35 and 164; Mann, "Chaos Theory and Strategic Thought"; and Waldman, *War, Clausewitz, and the Trinity*, 129.
22. Kahn, "Clausewitz and Intelligence," 117–126; and Sumida, *Decoding Clausewitz*, 128.
23. Clausewitz, *On War* (Howard and Paret), I.7, 120; and Clausewitz, *On War* (Howard and Paret), II.2, 143.
24. Howard, *Clausewitz*, 24.
25. Clausewitz, *On War* (Howard and Paret), I.3, 101.
26. Clausewitz, *On War* (Howard and Paret), I.3, 136. See also Stoker, *Clausewitz: His Life and Work*, 270.
27. Clausewitz, *On War* (Howard and Paret), I.3, 101; and Clausewitz, *On War* (Howard and Paret), II.2, 140. See also Waldman, *War, Clausewitz, and the Trinity*, 116.
28. See, for example, Brigety, *Ethics, Technology, and the American Way of War*, and Owens, *Lifting the Fog of War*.
29. Clausewitz, *On War* (Howard and Paret), I.3, 102. See also Waldman, *War, Clausewitz, and the Trinity*, 118.
30. For example, advances in technology allow senior commanders the proverbial "5000-mile screwdriver" to direct operations from afar, introducing additional layers of complexity and friction for the war fighter. For an early example of this dynamic, see Bowden, *Black Hawk Down*.
31. There is considerable nuance in the translation of the terms "politics" and "continuation," which remains the subject of academic debate to this day. Bassford, *Clausewitz in English*, 22.
32. While Tuchman is more nuanced than some of her contemporaries, she nevertheless falls into the trap of portraying Clausewitz as a driver of German strategic thinking on the eve of the Great War. Tuchman, *The Guns of August*, 13, 21, 24, 26, 36, 38, and 350–351. See also Münkler, *The New Wars* ; Kaldor, *New & Old War*, Holsti, *War, the State, and the State of War*; and Keegan, *A History of Warfare*.
33. Echevarria, *Clausewitz and Contemporary War*, 85–86; and Strachan, *Clausewitz's On War*, 147 and 152.

34. Clausewitz, *On War* (Howard and Paret), I.1, 87

35. By 1827 Clausewitz decided upon this as the unifying theme of his book and incorporated this insight, albeit imperfectly, into subsequent revisions. Strachan, *Clausewitz's On War*, 78 and 85.

36. Clausewitz, *On War* (Howard and Paret), I.1, 86.

37. Lebow, *The Tragic Vision of Politics*, 189–191.

38. Clausewitz, *On War* (Howard and Paret), I.1, 87.

39. Clausewitz, *On War* (Howard and Paret), I.1, 579.

40. Clausewitz, *On War* (Howard and Paret), VIII.3, 593. See also Paret, *The Cognitive Challenge of War*, 7.

41. Clausewitz, *On War* (Howard and Paret), I.1, 75.

42. Clausewitz.

43. Clausewitz, 77.

44. Bernhardi, *Germany and the Next War*.

45. Clausewitz, *On War* (Howard and Paret), I.1, 75–77

46. Clausewitz, *On War* (Howard and Paret), I.1, 78–80. See also Herberg-Rothe, *Clausewitz's Puzzle*, 53; and Sumida, *Decoding Clausewitz*, 122.

47. Clausewitz, *On War* (Howard and Paret), I.1, 80.

48. Clausewitz.

49. Clausewitz, 87.

50. Aron, *Clausewitz: Philosopher of War*, 100–101.

51. Clausewitz, *On War* (Howard and Paret), I.1, 75–77. See also Sumida, *Decoding Clausewitz*, 121.

52. Clausewitz, *On War* (Howard and Paret), I.2, 92.

53. Clausewitz. See also Daniel and Smith, "Burke and Clausewitz on the Limitation of War," 313–330.

54. Clausewitz, *On War* (Howard and Paret), VIII.3, 585.

55. Clausewitz.

56. Clausewitz's choice of the term *Dreifahltigkeit* for "trinity" appears deliberate, as it evokes the three-in-one mystery of the Christian faith. Echevarria, *Clausewitz and Contemporary War*, 70; and Herberg-Rothe, *Clausewitz's Puzzle*, 98.

57. Coker, *Barbarous Philosophers*, 87.

58. Daniel and Smith, "Burke and Clausewitz on the Limitation of War," 72–73.

59. There is also a striking similarity in this early work to Machiavelli's view of fortune. Stoker, *Clausewitz: His Life and Work*, 88.

60. Clausewitz, *On War* (Howard and Paret), I.1, 89.

61. Clausewitz.
62. Clausewitz appears to have been making the same point that Hegel did—that war is waged by individuals on behalf of the state. Strachan, *Clausewitz's On War*, 91.
63. Fleming, *Clausewitz's Timeless Trinity*, 49–57; and Waldman, *War, Clausewitz, and the Trinity*, 166.
64. Some scholars have attributed this misunderstanding, at least in part, to the translation by Howard and Paret. Fleming, *Clausewitz's Timeless Trinity*, 49–57.
65. Handel, *Masters of War*, 81.
66. Fleming, *Clausewitz's Timeless Trinity*, 2 and 54.
67. Many late twentieth-century critics have privileged the rational element of government over the other parts of the trinity—a reflection of their own historical context and perhaps the destructive potential of nuclear weapons—but this constitutes a fundamental misreading of Clausewitz's theory. Strachan, *Clausewitz's On War*, 178–179.
68. Clausewitz, *On War* (Howard and Paret), I.1, 89.
69. Clausewitz, 89.
70. Waldman, *War, Clausewitz, and the Trinity*, 161 and 174–175.
71. Clausewitz may not have believed that the trinity could be fully understood, but he saw it as a way to convey a deeper truth about the nature of war. Fleming, *Clausewitz's Timeless Trinity*, 2 and 185.
72. For the parallels between Hegel and Clausewitz's thoughts on this point, see Strachan, *Clausewitz's On War*, 91.

CHAPTER 8

THE SHIFTING
CHARACTER OF WAR

In many ways, Clausewitz lived and wrote at an inflection point in military affairs.[1] During his lifetime, he witnessed the demise of Frederick the Great's paradigm of limited wars; the rise of nationalism and "people's wars"; the emergence of modern bureaucratic nation states; improvements in logistics; the development of mass armies; and numerous tactical innovations, including skirmishing, light infantry tactics, and improved field artillery. While Clausewitz did not believe that these developments altered the nature of war, he clearly believed that they had transformed its character.

To describe these changes, Clausewitz developed four concepts to explain the transformation of warfare during his time—particularly the mass, ideologically driven wars of the French Revolutionary and Napoleonic era. First, people's wars were uniquely destructive and protracted. Second, defense was the strongest form of warfare. Third, the best chance of success lay in striking an opponent's center of gravity. Fourth, once commanders had reached the culminating point of an attack, they should halt offensive operations. In each of these concepts,

Clausewitz sought to make sense of the transformations he had witnessed over the course of his career, and to offer guidance to military practitioners confronting the new and increasingly lethal warfare.

People's Wars

The French Revolution and the subsequent Napoleonic Wars unleashed a massive amount of pent-up energy. Revolutionary ideas of liberty, equality, and fraternity were more than a challenge to the *Ancien Régime* of France—they transformed the political and security landscape of Europe.[2] Clausewitz witnessed these revolutions in military affairs firsthand during his campaigns along the Rhine and in Russia, and he followed with great interest the unfolding guerrilla campaigns against Napoleon on the Iberian Peninsula.[3]

Clausewitz recognized that although the particulars of these campaigns differed greatly, the underlying drivers were the same. The passions of the people proved to be a remarkably powerful motivation to fight, transforming both the ends for which wars were fought and the means by which they were conducted.[4] Wars were no longer fought between monarchs for limited aims of territory or prestige. Instead, they were waged by masses of people for transformative ends—namely, determining the geopolitical fate of entire nations.

This expansion of ends corresponded with an expansion of means. The small professional armies of the past, which had numbered in the tens of thousands, were quickly replaced by the mass armies of citizen-soldiers numbering in the hundreds of thousands or even millions. What these forces lacked in military efficiency, they more than compensated for in sheer numbers. As a result, they could sustain and inflict levels of casualties previously unimaginable, all in pursuit of their transformative ends.

While many of the monarchies of Europe were slow to respond to the changing character of war, Clausewitz recognized the critical importance of these developments. He dedicated significant space in *On War* to documenting the history of these changes and building theories to better understand them. As he observed, "at the outbreak of the French Revolution," war was seen as limited by the European powers, who responded to the French with limited means:

> Austria and Prussia tried to meet this with the diplomatic type of war that we have described. They soon discovered its inadequacy. Looking at the situation in this conventional manner, people at first expected to have to deal only with a seriously weakened French army; but in 1793 a force appeared that beggared all imagination.

The force that so shocked the Austrians and Prussians represented something fundamentally different. By harnessing the energies of the people, backing mass armies with the resources of modern nation-states, and applying modern scientific methods, these armies became almost unrecognizable. Clausewitz continues:

> Suddenly war became the business of the people—a people of thirty millions, all of whom considered themselves to be citizens...The people became a participant in war; instead of governments and armies as heretofore, the full weight of the nation was thrown into the balance. The resources and efforts now available for use surpassed all conventional limits; nothing now impeded the vigor with which war could be waged, and consequently the opponents of France faced the utmost peril....this juggernaut of war, based on the strength of the entire people, began its pulverizing course through Europe. It moved with such confidence and certainty that whenever it was opposed by armies of the traditional type there could never be a moment's doubt as to the result.[5]

The character of war had been transformed by this unleashed energy.[6]

Perhaps the starkest elucidation of this revolutionary energy was the 1793 declaration by the French National Assembly:

> The entire French nation is permanently called to the colors. The young men will go into battle; married men will forge weapons and transport supplies; women will make tents and uniforms and

serve in hospitals; children will make old cloth into bandages; old men will have themselves carried to the public squares to rouse the courage of the warriors and preach hatred of kings and unity of the Republic.[7]

As a result of this *levée en masse*, the size of the French army expanded dramatically—from approximately 155,000 in 1792 to over 700,000 by the close of 1793.[8] This was a radical shift. Not only were the French revolutionaries expanding their military, but they were also transforming the character of war into something fundamentally larger and more destructive.[9]

This rapid expansion necessitated the inclusion of entire sectors of the population that had previously been excluded from military service. One of the most radical aspects of the French Revolution was the incorporation of the middle classes into military ranks. Whereas the *Ancien Régime* had not required large numbers of recruits and had been content to draw from the lowest rungs of society, patriotic duty and practical necessity now demanded participation from both the middle and working classes.[10] This infusion of human capital aligned naturally with the revolutionary ideals of comradeship, shared sacrifice, and patriotism. Through the *levée en masse*, the revolutionary call of "liberty, egality, and fraternity" became a practical reality for the masses. Clausewitz described this transformation in striking terms:

> The revolutionary methods of the French had attacked the traditional ways of warfare like acid; they had freed the terrible element of war from its ancient diplomatic and economic bonds. Now war stepped forth in all its raw violence, dragging along an immense accumulation of power, and nothing met the eye but the ruins of the traditional art of war on one hand, and incredible successes on the other.[11]

Suddenly, military campaigns were much more than the sport of kings, conducted with small professional armies on the edges of a map. Now,

common people from all classes of society had a personal connection to
the outcome of these great campaigns.

As wars grew larger, their horrors began to increasingly impact
civilians in ways not seen in Europe since the Thirty Years' War (1618–
1648).[12] The new mass armies were so large that they were forced to
requisition food from local populations and quarter soldiers in civilian
homes. This naturally led to violence between military forces and the
civilians they encountered, expanding the scope of warfare in a new
and horrific way. "This method knew no limits other than the complete
exhaustion, impoverishment and devastation of the country."[13] While
Clausewitz did not live to see the mass murders of the twentieth century,
this warfare was more brutal than anything in Europe's living memory
and seemed to prove his theory that war tends toward extremes.

During the Napoleonic Wars, there were numerous technological
changes (e.g., canned food, improvements in field artillery, and semaphore
telegraphy), but Clausewitz minimized their importance. He instead
focused on the radical societal changes that had taken place during
this period. According to Clausewitz, very few changes in war "can be
ascribed to new inventions or new departures in ideas" rather, they arise
"mainly from the transformation of society and new social conditions."[14]
Many of Napoleon's opponents, including Clausewitz's native Prussia,
were slow to recognize this fundamental shift. At Jena, Auerstädt, and
dozens of other battlefields, Prussia and its allies fought by the old rules,
failing to grasp that "the colossal weight of the whole French people,
unhinged by political fanaticism, came crashing down [upon] us."[15]
Warfare had fundamentally changed and the "forms and means" of the
old Prussian system "were no longer appropriate to the changed times
and political conditions."[16]

In addition to these "big wars," guerrilla wars, insurgencies, and
popular resistance movements also proliferated during this period.[17]
Clausewitz lamented that people's resistance had spread to the "civilized
parts of Europe," but he also recognized that the rise of these conflicts

heralded a shift in the conduct of war. Despite Clausewitz's distaste of these "popular uprisings," he studied them with great interest as early as his time as an instructor at the war college in 1810. He clearly saw these wars were important and, at the time of his death, was planning to incorporate them more fully into *On War*.[18]

In Clausewitz's mind, the "phenomena of the nineteenth century" such as the Peninsula War and Russian scorched-earth policies, were logical responses to the power and ambition of Napoleon.[19] Although Napoleon had harnessed the power of the people to wage conventional war, that same force ultimately contributed to his downfall. By rousing the energies of the people and compelling them to join insurgent movements, Napoleon unleased a form of resistance for which he was unprepared. These guerrilla forces did not fight set-piece battles, instead they "nibble around the shell and around the edges." By avoiding pitched battles and maintaining a persistent presence, they "arouse uneasiness and fear and deepen the psychological effect of the insurrection as a whole."[20] Through such tactics, irregular forces can erode an adversary's will to fight and attain political goals despite being militarily weaker.

People's wars prolonged conflicts and increased their destructiveness by enabling nations to continue resisting an invader even after the defeat of their conventional forces. In such contexts, decisive victory may be impossible as long as a nation is willing to sustain resistance:

> A government must never assume that its country's fate, its whole existence, hangs on the outcome of a single battle, no matter how decisive. Even after a defeat, there is always the possibility that a turn of fortune can be brought about by developing new sources of internal strength or through the natural decimation that all offensives suffer in the long run or by means of help from abroad. There will always be time enough to die; like a drowning man who will clutch instinctively at a straw, it is the natural law of the moral world that a nation that finds itself on the brink of an abyss will try to save itself by any means.[21]

An impassioned people thus possessed both the power and the duty to resist: "No matter how small and weak a state may be in comparison with its enemy, it must not forgo these last efforts, or one would conclude that its soul is dead."[22]

By equating a nation's continued willingness to resist after suffering a military defeat with its moral strength, Clausewitz implicitly criticized his sovereign, Frederick Wilhelm III, who had refused to authorize guerrilla warfare in 1806 and again in 1812.[23] Clausewitz continues:

> A government that, after having lost a major battle, is only interested in letting its people go back to sleep in peace as soon as possible, and, overwhelmed by feelings of failure and disappointment, lacks the courage and desire to put forth a final effort, is, because of its weakness, involved in a major inconsistency in any case. It shows that it did not deserve to win, and, possibly for that very reason, was unable to.[24]

For Clausewitz, the passions of the people and their governments' commitment to warfare had fundamentally transformed war itself, making it far more destructive and protracted.[25] He observed that future wars would be more destructive, "a consequence of the way in which our day the elemental violence of war has burst its old artificial barriers."[26] The fusion of popular energy and national endurance led Clausewitz to his next point: the primacy of the defense.

Primacy of the Defense

Despite war's tendency toward increasingly violent extremes, Clausewitz believed that conquest would remain difficult to achieve. His reasoning was straightforward: although the energy of the people could intensify warfare, it also posed challenges for would-be conquerors. Equally important was Clausewitz's conviction that the defense was the stronger form of warfare. This insight was central to his theory, allowing him to reconcile the concepts of friction, shifting fortunes, and the difficulty of achieving decisive victory in an age of people's wars.

Clausewitz's views on the primacy of the defense can be traced as far back as 1804 when he read Machiavelli's *Discourses* and composed a short essay on it.[27] Machiavelli praised the Roman general Quintus Fabius Maximus Verrucosus —known as the Cunctator, or "the delayer"—who used delaying tactics to wear down Hannibal's forces during the Second Punic War. Although the Roman public eventually rejected these "Fabian" tactics and demanded direct engagement, Machiavelli argued they would have succeeded had they not been abandoned under political pressure. Drawing on this example, Clausewitz began to formulate his view that the defense , while inherently passive, was the stronger form of warfare.[28]

Prior to joining the Russian army in 1812, Clausewitz expanded upon these ideas in an addendum to the unpublished political writings.[29] While he acknowledged that France was stronger than Prussia in conventional terms, he argued that the Prussians could prevail because they would be fighting defensively and for their fatherland, whereas their French opponents would face the more difficult task of suppressing a rebellion.[30] Although the hoped-for revolt never materialized, Clausewitz was clearly already developing his theory on the primacy of defensive warfare.

Clausewitz's service in the 1812 Russian campaign added a powerful layer of personal observation to his earlier insights. While serving in the tsar's army, he witnessed Napoleon's inability to achieve a rapid victory. As the French advanced deeper into Russia, their strength waned. The Russian forces traded space for time and avoided giving Napoleon the decisive battle he sought.[31] The Russians followed what Clausewitz described as a strategy of "voluntary withdrawal to the interior of the country...[which] destroys the enemy not so much by the sword as by his own exertions."[32] Even though Napoleon held the field after the Battle of Borodino, "the 1812 campaign failed because the Russian government kept its nerve and the people remained loyal and steadfast.[33]" Napoleon entered a burned and abandoned Moscow and found himself unable to continue the campaign. The onset of winter, the collapse of his supply lines, and determined Russian counterattacks combined to bring about the destruction of Napoleon's army.[34] Clausewitz later asserted that while no one would have criticized Napoleon's campaign had it succeeded, its failure underscored the difficulty of achieving ambitious political ends through military action.[35]

By the time he began composing *On War*, Clausewitz was convinced that the defense was the strongest form of warfare. Although Clausewitz reserved the majority of his discussion on this point for Books VI and VII, he previewed it in the opening lines of Book I, Chapter 1.[36] There, he quickly refutes the idea that wars can be won with a "single, short blow" and observes that even when "the balance has been badly upset, equilibrium can be restored."[37] He writes:

> As we shall show defense is a stronger form of fighting than attack...I am convinced that the superiority of the defensive (if rightly understood) is very great, far greater than appears at first sight. It is this which explains without any inconsistency most periods of inaction that occur in war. The weaker the motives for action, the more will they be overlaid and neutralized by this disparity between attack and defense, and the more frequently will action be suspended—as indeed experience shows.[38]

In these opening passages, Clausewitz ties the primacy of the defense to themes that he further develops in later chapters—namely, friction and political aims. The defense can prevail over stronger forces by "wearing down" the attacker and "using the duration of the war to bring about the gradual exhaustion of his physical and moral resistance."[39]

Delays or failures to achieve decisive results erode the aggressor's political will. The defender, by contrast, pursues more limited and attainable political ends:

> The minimum object is *pure self-defense*; in other words, fighting without a positive purpose. With such a policy our relative strength will be at its height, and thus the prospects for a favourable outcome will be the greatest (emphasis in the original).[40]

Success on the defensive could enable the defender to transition to the offensive and pursue the larger goal of destroying the former aggressor:

> This usually means that *action is postponed* in time and space to the extent that space and circumstances permit....The destruction of the enemy—an aim that has until then been postponed but not displaced by another consideration—now reemerges (emphasis in the original).[41]

Such a shift would reverse the roles of the two belligerents. The former attacker, now on the defensive, would regain the advantage, while the former defender—now the attacker—would face the challenges of pursuing more ambitious political ends. Friction and the difficulty of achieving positive ends would again favor the defender, perpetuating the cycle. A successful strategist must anticipate and account for these shifting dynamics in the planning and conduct of war.

While Book I contains the clearest statement of the primacy of the defense, Clausewitz clearly intended to incorporate this insight throughout his work. In his unpublished 1830 note, Clausewitz wrote that he hoped to expand on his observation that "the defense is the stronger form of fighting with the negative purpose, attack the weaker form with the positive purpose" throughout his unfinished manuscript.[42] Clausewitz's commitment to the theoretical importance of defensive warfare is also evident in the fact that Book VI, titled "Defense," is more than twice the length of the next longest book (Book V), and three times the length of most others (Books I, II, III, and VII).[43] Had Clausewitz lived, he might have shortened Book VI or more fully integrated this theory throughout the body of his text; however, given the unfinished nature of the work, such speculation remains uncertain.[44] What is clear, however, is that Clausewitz saw defensive warfare as a foundational concept and addressed it at length, especially in Book VI.

In Book VI, Clausewitz expands on his insight that defense, as a reactive form of warfare, consists of two basic elements, "the parrying of a blow" and "awaiting the blow."[45] The fact that the defender can wait to exploit the actions of the attacker provides a critical advantage:

> [T]ime which is allowed to pass unused accumulates to the credit of the defender. He reaps what he did not sow. Any omission of attack—whether from bad judgement, fear, or indolence—accrues to the defender's benefit.[46]

While Clausewitz is clear that "defense is the stronger form of war," he recognizes it "has a negative object" and is typically inadequate for achieving political ends beyond national survival.[47]

Since defense has only a negative purpose, even the defending power must eventually prepare for offensive operations if it is to achieve its aims:

> it follows that it should be used only so long as weakness compels, and be abandoned as soon as we are strong enough to pursue a positive object. When one has used defensive measures successfully, a more favorable balance of strength is usually created; thus, the natural course in war is to begin defensively and end by attacking...In other words, a war in which victories were used only defensively without the intention of counterattacking would be absurd as a battle in which the principle of absolute defense—passivity, that is—were to dictate every action.[48]

This creates a dynamic in which the advantage in war can shift rapidly as one side transitions offense to defense. In this fluid system, "War serves the purpose of the defense more than that of the aggressor."[49] Since time and distance work against the attacker, they are compelled to win quickly and decisively before their strength is eroded. Thus, "The Aggressor is always peace-loving (as Bonaparte claimed to be)."[50]

Given the shifting fortunes of war, once an attacker has exhausted its strength, the defender should rapidly exploit the advantage and counterattack:

> Once the defender has gained an important advantage, defense as such has done its work. While he is enjoying this advantage,

he must strike back, or he will court destruction. Prudence bids him strike while the iron is hot and use his advantage to prevent a second onslaught...This transition to the counterattack must be accepted as a tendency inherent in defense—indeed, as one of its central features.[51]

Clausewitz may have had in mind the failure to destroy Napoleon during his retreat from Russia in 1812 when he wrote:

Whenever a victory achieved by the defensive form is not turned to military account, where, so to speak, it is allowed to wither away unused, a serious mistake is being made. A sudden powerful transition to the offensive—the flashing sword of vengeance—is the greatest moment for the defense.[52]

Recognizing that many strategists favor the offensive, Clausewitz argues that prudent commanders should not dismiss defensive strategies as weak or indecisive:

...the general will let the enemy come on, not from confused indecision and fear, but by his own choice, coolly and deliberately; fortresses are undaunted by the prospect of a siege, and finally a stout-hearted populace is not more afraid of the enemy than he of it. This constituted, defense will no longer look so easy and infallible as it does in the gloomy imagination of those who see courage, determination, and movement in attack alone, and in defense only impotence and paralysis.[53]

Clausewitz continues this argument in Book VII, which focuses on the attack, where he identifies seven factors that cause offensive strength to diminish:

The diminishing force of the attack is one of the strategist's main concerns. His awareness of it will determine the accuracy of his estimate in each case of the options open to him. Overall strength is depleted:

1. If the object of the attack is to occupy the enemy's country (Occupation normally begins only after the first decisive action, but the attack does not cease with this action.)"

2. By the invading armies' need to occupy the area in their rear so as to secure their lines of communication and exploit its resources

3. By losses incurred in action and through sickness

4. By the distance from the source of replacements

5. By sieges and the investment of fortresses

6. By a relaxation of effort

7. By the defection of allies.[54]

By placing this warning in the section about attack, Clausewitz reinforced his conviction that the defense is the stronger form of warfare.

For Clausewitz, the superiority of the defense was not merely a theoretical proposition; it was a practical lesson for strategists. Wars are not easily won, and achieving political ends is exceedingly difficult. Strategists must therefore adopt a cautious approach that acknowledges these enduring constraints.

Center of Gravity

Throughout *On War*, Clausewitz emphasized that the wars of his time were becoming large in scale, extremely destructive, and difficult to win. Although Clausewitz advised extreme caution, he did not believe that victory was impossible. To reconcile this tension between caution and the possibility of victory, Clausewitz developed the concept of the center of gravity. This theory was so important that Clausewitz refers to it more than fifty times in *On War*, underscoring its central role in his theory of war.[55]

Yet, despite Clausewitz's clear intention to provide a theory for achieving victory, his treatment of the concept is somewhat inconsistent—likely due to the unfinished nature of *On War*. At times, he presents it in tactical terms; at others, as a strategic consideration. He also appears uncertain about how to define center of gravity or how to identify it. Is it the enemy's army? Its lines of communication? A capital city or a major industrial hub? Clausewitz acknowledges that center of gravity is variable and sometimes difficult to discern. Given these internal inconsistencies, it is essential to approach this concept with caution and an awareness of its limitations.

In its simplest form, center of gravity refers to the point from which a belligerent derives the strength necessary to sustain resistance. This critical source of strength is, therefore, also the point of greatest vulnerability.[56] From this observation, it flows logically that a commander seeking victory should concentrate maximum effort against the enemy's center of gravity while safeguarding their own.[57]

The concept of center of gravity can be applied in either a tactical or a strategic context. In the tactical sense, Clausewitz used it to describe the point on the battlefield where an enemy is most vulnerable. This is similar to the Jominian concept of directing maximum force at a critical point. Clausewitz writes that the tactical center of gravity "is always found where the mass is concentrated most densely."[58] At the strategic level, center of gravity refers to the element that, if destroyed or neutralized, would compel an opponent to abandon resistance. Clausewitz believed that the "real key to an enemy's country is usually his army," but center of gravity could also be a capital city, a key geographic position, an economic hub, the defection of an ally, or any other factor capable of forcing capitulation. In short, center of gravity was central to Clausewitz's thinking—but it was not a single, fixed concept.

For many, the fact that center of gravity is not a single, identifiable element—such as a nation's capital—is deeply dissatisfying. While strategists often desire a clear formula for success (e.g., seize the capital to win the war), Clausewitz offers no such recipe. This is not the result of flawed reasoning, but rather a reflection of Clausewitz's probabilistic approach to war and his historically informed perspective. During the Napoleonic Wars, Clausewitz observed firsthand how center of gravity could shift over time. In 1805, for example, enemy armies served as the center of gravity. Napoleon secured decisive victories at Ulm and Austerlitz by destroying his opponents' forces and compelling them to submit. In 1806–1807, the decisive factor was his enemies' political fragmentation. By moving rapidly and defeating them separately, Napoleon prevented their unification into a coherent alliance. In 1812, however, he discovered to his dismay that neither battlefield victories nor territorial gains could bring decisive success. Because the true center of gravity was the will of the Russian people—something not easily broken—Napoleon's campaign unraveled on the long road to Moscow.[59]

Ultimately, Clausewitz's theory compels the strategist to identify their opponent's center of gravity and to develop "the plan of war designed to lead to a total defeat of the enemy."[60] While not easy to employ, the concept aligns closely with Clausewitz's broader philosophical approach to war. The trinity's constant tension ensures that no center of gravity remains fixed or permanent. In such a world, the ability to grasp this fluid and complex dynamic marks the true master strategist. Yet even a clear understanding of an opponent's center of gravity—and the means to strike it—does not guarantee victory.

Culminating Point of the Attack
While Clausewitz warned that defense was stronger and encouraged that maximum effort be directed at an opponent's center of gravity, he understood that this was insufficient guidance for aspiring strategists. To make his theory of war more applicable, he also needed to consider what to do when victory proved impossible. Indeed, knowing when to stop and accept limited gains—or to limit future losses—was an essential skill to master.

To this end, Clausewitz developed the concept of the culminating point of the attack. In its simplest form, this refers to the place or moment where, once achieved, the campaign might be concluded, but beyond which the attacker's efforts become counterproductive. Clausewitz describes this concept as:

> There are strategic attacks that have led directly to peace, but these are the minority. Most of them only lead up to a point where their remaining strength is just enough to maintain a defense and wait for peace. Beyond this point the scale turns and the reaction follows with a force that is usually much stronger than that of the original attack. This is what we mean by the culminating point of the attack.[61]

Because of the durability and resilience of modern nation-states, Clausewitz understood that it was almost impossible to fully conquer or destroy an opponent.

This posed the ultimate Goldilocks problem—how to win a decisive battlefield victory and secure a satisfactory peace without provoking the enemy to fight even harder or the people to rise in rebellion. Clausewitz described this dilemma facing a general who:

> must *guess*, so to speak: guess whether the first shock of battle will steel the enemy's resolve and stiffen his resistance, or whether, like a Bologna flask, it will shatter as soon as its surface is scratched; guess the extent of deliberation and paralysis that the drying up of particular sources of supply and the severing of certain lines of communication will cause in the enemy; guess whether the burning pain of injury he has dealt will make the enemy collapse with exhaustion or, like a wounded bull, arouse his rage; guess whether the other powers will be frightened or indignant, and whether and which political alliances will be dissolved or formed.[62]

Clausewitz had experienced this problem firsthand on countless battlefields and was keenly aware that it posed an incredibly difficult decision for a commander who is presented with "thousands of wrong turns running in all directions."[63] Clausewitz continues this sympathetic understanding:

> If we remember how many factors contribute to an equation of forces, we will understand how difficult it is in some cases to determine which side has the upper hand. Often it is entirely a matter of the imagination. What matters therefore is to detect the culminating point with discriminative judgement. We here come up against an apparent contradiction. If defense is more effective than attack, one would think that the latter could never lead too far; if the less effective form is strong enough the more effective form should be even stronger.[64]

Given this complexity, Clausewitz notes that many generals err on the side of caution: "This is why the great majority of generals will prefer to stop well short of their objective rather than risk approaching it too closely," while others "with high courage and an enterprising spirit will often overshoot it and so fail to attain their purpose."[65] Although Clausewitz understood why commanders might be too cautious, he believed this was as much of a mistake as being too aggressive. Unfortunately, Clausewitz never defined when or how a campaign has reached its productive limit and instead placed the burden on the commander to act wisely.

Given this tension, Clausewitz's theory of the culminating point of the attack once again presents the reader with a somewhat unsatisfying set of theoretical extremes, not a simple answer. Rather than clearly state when a commander has reached the culminating point of victory, Clausewitz urged commanders to use imagination and judgment, and to embrace uncertainty and contradictions. While this approach is entirely consistent with the broader probabilistic approach of *On War*, it is more a philosophy than a recipe for success.

This philosophy of war asks a lot of a modern commander. The commander must understand that people's wars are uniquely destructive. In an era where a decisive and rapid victory is increasingly difficult to achieve, the commander must still strive to attain meaningful results. The commander must correctly identify and destroy an opponent's center of gravity while preserving their own. The commander must understand when their attacks have reached their limits and restrain themselves accordingly. In a world of friction, uncertainty, and probabilities, this requires countless virtues including wisdom, knowledge, luck, and perseverance. In short, it requires a commander with military genius. The concept of genius is one of the most important and enduring elements of Clausewitz's theory and is the subject of the next chapter.[66]

NOTES

1. Lebow, *The Tragic Vision of Politics*, 176.
2. Bell, *The First Total War*, and Lebow, *The Tragic Vision of Politics*, 168.
3. Sumida, *Decoding Clausewitz*, 164.
4. As economic historian Niall Fergusson noted, "Marx had famously called religion the 'opium of the masses.' If so, then nationalism was the cocaine of the middle classes." Ferguson, *Civilization*, 242. See also Hirschman, *The Passions and the Interests*, 61, 113, 119–120.
5. Clausewitz, *On War* (Howard and Paret), VIII.3, 591–593.
6. Bell, *The First Total War*.
7. Knox and Murray, *The Dynamics of Military Revolution, 1300–2050*, 8. For a historical perspective on this revolution in military affairs, see McPhee, *Liberty or Death*, 210–211.
8. McPhee, *Liberty or Death*, 210–211; and Willmott and Barrett, *Clausewitz Reconsidered*, 29. After the defeats in Russia and at Leipzig, a similarly ambitious *levée en masse* raised nearly a million new troops for Napoleon's army. Esdaile, *Napoleon's Wars*, 519.
9. Stoker, *Clausewitz: His Life and Work*, 18–22.
10. Howard, *The Invention of Peace*, 33–60.
11. Clausewitz, "On the Life and Character of Scharnhorst," in Clausewitz, *Carl von Clausewitz: Historical and Political Writings*, 102.
12. Parker, *Global Crisis*.
13. Clausewitz, *On War* (Howard and Paret), V.14, 336. See also Strachan, *Clausewitz's On War*, 121.
14. Clausewitz, *On War* (Howard and Paret), VI.30, 515. See also Strachan, *Clausewitz's On War*, 36–37.
15. Clausewitz, *On War* (Howard and Paret), VI.30, 518.
16. Clausewitz, "On the Life and Character of Scharnhorst" in Clausewitz, *Carl von Clausewitz: Historical and Political Writings*, 102.
17. Clausewitz, *Clausewitz On Small War*. See also Echevarria, *Clausewitz and Contemporary War*, and Scheipers, *On Small War*.
18. Clausewitz, *On War* (Howard and Paret), VI.26, 479.
19. Clausewitz. See also Esdaile, *The Peninsular War*; Sumida, *Decoding Clausewitz*, 164; and Lieven, *Russia Against Napoleon*.
20. Clausewitz, *On War* (Howard and Paret), VI.26, 479–483. See also Lebow, *The Tragic Vision of Politics*, 193.

21. Clausewitz, *On War* (Howard and Paret), VI.26, 483.
22. Clausewitz.
23. Sumida, *Decoding Clausewitz*, 165.
24. Clausewitz, *On War* (Howard and Paret), VI.26, 483.
25. Coker, *Barbarous Philosophers*, 88; and Daniel and Smith, "Burke and Clausewitz on the Limitation of War." According to one prominent scholar, Clausewitz had reached this conclusion by 1827, as evidenced by a memorandum from that year. Strachan, *Clausewitz's On War*, 189–191.
26. Clausewitz, *On War* (Howard and Paret), VI.26, 481.
27. Interestingly, Clausewitz's 1804 essay is quoted almost word for word in *On War*. Heuser, *Reading Clausewitz*, 91–92.
28. Paret, *Clausewitz and the State*, 169–171.
29. Stoker, *Clausewitz: His Life and Work*, 90 and 96–98.
30. Clausewitz, *Clausewitz On Small War*, 209–216. See also Hagemann, *Revisiting Prussia's Wars Against Napoleon*, 133.
31. See Clausewitz's discussions about the preservation of the army being more important than the preservation of territory: Clausewitz, *On War* (Howard and Paret), VI.27, 485.
32. Clausewitz, 469.
33. Clausewitz, 628.
34. Sumida, *Decoding Clausewitz*, 163–164.
35. Clausewitz, *On War* (Howard and Paret), VIII.9, 627–628.
36. Sumida, *Decoding Clausewitz*, 153.
37. Clausewitz, *On War* (Howard and Paret), I.1, 79. According to one scholar, the origins of this insight can be traced as far back as 1811 in a letter to Gneisenau. See Stoker, *Clausewitz: His Life and Work*, 90.
38. Clausewitz, *On War* (Howard and Paret), I.1, 84.
39. Clausewitz, 90–91
40. Clausewitz, 99.
41. Clausewitz.
42. Clausewitz, "Unfinished Note, Presumably Written in 1830," in Clausewitz, *On War* (Howard and Paret), 71. See also Sumida, *Decoding Clausewitz*, 155–156.
43. Sumida, *Decoding Clausewitz*, 156–157.
44. Indeed, Book VI is the most repetitive and most incomplete of Clausewitz's books, while Book VII is the least developed and shortest. See Strachan, *Clausewitz's On War*, 80–81 and 161. Raymond Aron viewed Book VI as a critical theoretical element of Clausewitz's thought and speculated that had Clausewitz lived to complete the work he would

have made the primacy of the defense the foundation for his theory of conflict resolution. See Aron, *Clausewitz: Philosopher of War*, 145–171.

45. Clausewitz, *On War* (Howard and Paret), VI.1, 357.
46. Clausewitz, 357.
47. Clausewitz.
48. Clausewitz, 358.
49. Clausewitz, 370.
50. Clausewitz.
51. Clausewitz.
52. Clausewitz. See also Sumida, *Decoding Clausewitz*, 163–164.
53. Clausewitz, *On War* (Howard and Paret), VI.5, 371.
54. Clausewitz 527. See also Echevarria, *Clausewitz and Contemporary War*, 161.
55. Echevarria, *Clausewitz and Contemporary War*, 177. See also Echevarria, "Clausewitz's Center of Gravity."
56. Clausewitz, *On War* (Howard and Paret), VI.28, 491.
57. Stoker, *Clausewitz: His Life and Work*, 272.
58. Clausewitz, *On War* (Howard and Paret), VI.27, 485. See also Strachan, *Clausewitz's On War*, 132.
59. Strachan, *Clausewitz's On War*, 133.
60. Clausewitz, *On War* (Howard and Paret), VIII.9, 621.
61. Clausewitz, 528.
62. Emphasis in original; Clausewitz, 572–573.
63. Clausewitz, 573.
64. Clausewitz, 528.
65. Clausewitz, 573.
66. Some would argue that Clausewitz's concept of genius is his single most important theoretical contribution. See Sumida, *Decoding Clausewitz*, 6.

CHAPTER 9

MILITARY GENIUS

Napoleon and the Shaping of Clausewitz's Thought

No single person shaped Clausewitz's career as much as Napoleon.[1]
Indeed, the French general was both his military nemesis and his intellectual inspiration. Clausewitz quite literally fought in dozens of battles,
endured imprisonment, defected from his country, and wrote hundreds
of pages—all because of this one man.

For Clausewitz, the impact of Napoleon's military genius was strikingly obvious—through his force of will, Napoleon literally defined the
Napoleonic Era. Napoleon's "audacity and luck" allowed him to break
free from the constraints that bound most other leaders and to "cast
the old accepted practices to the winds."[2] Typical rules and restraints
appeared not to apply to Napoleon, as he had the "talent and genius [to]
operate outside the rules" and operate in a unique space where, "theory
conflicts with practice."[3] Napoleon so thoroughly mastered the art of
war that he "achieve[d] [a] state of absolute perfection" and "[whose]
superiority has consistently led to the enemy's collapse." Indeed, he was
"to put it bluntly—the God of War himself." [4] For Clausewitz, this was
very high praise for a hated enemy.

Despite Napoleon's enormous impact, Clausewitz believed that Napoleon's influence on the conduct of war was not properly understood.[5] The initial response to Napoleon's revolutions in warfare was imitation, not a serious attempt at understanding. As a junior officer, Clausewitz participated in reform efforts and helped the Prussian army adopt tactical formations modeled on the French. Even then, Clausewitz regarded these reforms as long overdue but ultimately only a stopgap measure. It was not enough to copy tactical formations or adopt technical advances. One had to understand what made Napoleon different.

For Clausewitz, Napoleon provoked a visceral mix of emotions, including fear, anger, contempt, awe, and inspiration. He admired Napoleon as an unequaled professional soldier, yet as a Prussian nationalist, Clausewitz loathed him for the destruction he had wrought.[6] He recognized that Napoleon had transformed war in ways that made him uncomfortable, but he also understood that prudence demanded careful study of his campaigns. He struggled to understand how Napoleon could be both the greatest general of modern times and deeply flawed, even self-destructive. Ultimately, Clausewitz was able to compartmentalize his views, condemning Napoleon's character while praising his military abilities.[7] As such, it is essential to remember that Clausewitz's view of military genius is shaped by the complex character of Napoleon, the "God of War."

A Military Genius Must Have Sound Judgement

According to Clausewitz, military genius requires an uncommon mix of virtues including "a degree of intellectual powers," physical courage "in the face of personal danger," moral courage "to accept responsibility," physical strength, and the "powers of intellect" not be overcome by chance and uncertainty.[8] While all of these qualities were necessary, Clausewitz believed that the capacity to resist being overwhelmed by uncertainty and chance was paramount. Drawing from Immanuel Kant's *Critique of Judgement*, Clausewitz observed that most individuals struggle to make decisions based on imperfect or contradictory information.[9] Faced with this kind of uncertainty, many commanders become paralyzed by fear and doubt, often delaying action while awaiting perfect clarity, thus allowing the perfect to become the enemy of the good.

Yet despite these formidable constraints, Clausewitz maintained that military genius could still prevail. By swiftly assessing the available information and exercising superior judgment, a commander could achieve a functional understanding of the situation and act decisively. Victory, Clausewitz argued, would go not to the commander who aimed for perfection, but to the one who was less imperfect than the adversary—and who was not incapacitated by fear and indecision.[10]

The Twin Habits of Mind

Clausewitz identified two mental faculties as essential for military genius: *coup d'œil* and determination. As in other areas of his theory, Clausewitz drew inspiration from Napoleon. While many commanders possessed isolated elements of genius, Napoleon's enduring success, Clausewitz argued, derived from his extraordinary capacity to assess complex situations rapidly and act with bold resolve. These two habits of mind enabled him to prevail in the face of uncertainty and friction, where others hesitated or faltered.[11]

Coup d'œil takes its name from the French for "strike of an eye" and describes the almost instantaneous understanding of a military situation.[12] Clausewitz defines this faculty as "the inward eye...[and] the quick recognition of truth that the mind would ordinarily miss or would perceive only after long study and reflection."[13] This ability to assess unfolding circumstances rapidly was crucial because "action can never be based on anything firmer than instinct, and sensing of truth."[14] In combat, a general would rarely have sufficient time or information to each a perfectly rational decision; instead, "the man of action must at times trust in the sensitive instinct of judgement, derived from his native intelligence and developed through reflection, which almost unconsciously hits upon the right course."[15]

To cultivate this habit of mind, Clausewitz argued that officers should study history. Rather than merely memorize names and dates, they should seek to develop a practical understanding of how past leaders made decisions under pressure:

> [I]t is useful to study history in connection with this subject, as with others. While there may be no system, and no mechanical way of recognizing the truth, the truth does exist. To recognize it one needs seasoned judgement and instinct born of long experience. While history may yield no formula, it does provide an *exercise for judgement* here as everywhere else [emphasis in the original].[16]

By compelling officers to engage deeply with recurring military problems, Clausewitz believed that it was possible to develop "insights, broad impressions, and flashes of intuition" necessary to act decisively under the strain of battle.[17]

He argued that no two battles or campaigns are ever the same, but a military genius could use their accumulated understanding of decision making to anticipate and respond more effectively to unfolding situations. This practical approach to studying war was precisely what Clausewitz had unsuccessfully sought to incorporate into the curriculum of the Prussian war college in 1819. Although he did not live to see it realized, his vision for applied learning would become a central feature of German military education and would later be credited for the German army's operational effectiveness from the Franco-Prussian War to World War II.[18]

Although he was not formally trained as a neuroscientist, Clausewitz nonetheless stumbled upon a fundamental truth about unconscious cognition now known as dual process theory.[19] As modern studies demonstrate, the human brain can make decisions in ways that are either "fast" or "slow." Fast decision making is instinctive, emotional, and intuitive whereas slow decision making is more analytical, logical, and deliberate.[20] Recent studies have analyzed the cognitive processing of individuals with high-risk jobs, such as firefighters and fighter pilots, and discovered that the most successful members of these groups can think and act quickly. When confronted with a life-threatening decision, such as how to extinguish a fire or maneuver an aircraft in combat, this fast thinking is almost completely instinctual. In these circumstances, the brain must respond with speed and decisiveness in order to ensure survival.[21]

Clausewitz seemed to understand this phenomenon intuitively and incorporated it into his conception of military genius. Much like a fire-fighter confronting a blaze or a pilot navigating a high-performance aircraft, a general in the midst of battle has little time to deliberate—they must act quickly and decisively. According to Clausewitz, "almost every-thing happens in war *through the hidden process of intuitive judgement.*"[22] While it is important not to overstate this observation, Clausewitz's *coup d'œil* closely parallels the fast mode of decision making described nearly a century and a half later in dual process theory.

For Clausewitz, *coup d'œil* was one of the defining traits of military genius:

> When it is all said and done, it really is the commander's *coup d'œil*, his ability to see things simply, to identify the whole business of war completely with himself, that is the essence of good generalship. Only if the mind works in this comprehensive fashion can it achieve the freedom it needs to dominate events and not be dominated by them.[23]

While this was a key habit of mind, Clausewitz maintained that it was not sufficient to achieve military success.

The second habit of mind that Clausewitz believed was essential to military genius was determination—the ability to act upon one's convictions. According to Clausewitz, determination required a type of mental fortitude comparable to physical courage:

> Determination in a single instance is an expression of courage; if it becomes characteristic, a mental habit...the role of determination is to limit the agonies of doubt and the perils of hesitation when the

> motives for action are inadequate...Determination which dispels doubt, is a quality that can be aroused only by the intellect, and by a specific cast of mind at that. More is required to create determination than a mere conjunction of superior insight with the appropriate emotions.[24]

For Clausewitz, risking life and limb required little intelligence; it was the determined mind that could fully grasp the dangers and still face them.

Much like *coup d'œil*, determination was, for Clausewitz, a manifestation of military genius that is distinct from pure intelligence and academic reasoning. On the contrary, many highly intelligent individuals failed to act decisively in moments of critical decision. In contrast, the true genius possessed a strength of mind that enabled them to confront and overcome the immense challenges of war. To underscore this point, Clausewitz observed, "We admire presence of mind in an apt repartee, as we admire quick thinking in the face of danger. Neither needs to be exceptional, so long as it meets the situation."[25]

A Warning about Napoleon's Genius

While Napoleon may have embodied military genius—indeed, he was for Clausewitz the "God of War"—it is crucial to remember that he ultimately failed. A combination of increasingly capable adversaries, unwinnable campaigns, and his own egotism brought about his downfall.[26] According to Clausewitz, Napoleon's genius failed him because he did not recognize that:

> Appropriate talent is needed at all levels if distinguished service is to be performed...To bring a war, or one of its campaigns, to a successful close requires a thorough grasp of national policy. On that level strategy and policy coalesce: the commander-in-chief is simultaneously a statesman.[27]

As Clausewitz's theories predict, Napoleon's military genius enabled him to win dazzling battlefield victories, but it could not deliver unrealistic political ends.[28]

The Russian campaign of 1812 and the Hundred Days both serve as vivid illustrations of Napoleon's limitations. In Russia, Napoleon's genius failed him; he underestimated both the resilience of his opponent and the enormity of the task before him. Expecting to force a decisive battle, Napoleon was unprepared for Russia's strategy of continual retreat and scorched-earth resistance. He also failed to grasp the strategic implications of Russia's vast geography. Within such an expansive terrain, he could neither achieve a decisive victory through maneuver nor advance rapidly toward Moscow, even when it lay undefended.[29] These combined miscalculations led directly to catastrophe and exemplified how even a commander of Napoleon's stature could falter. In Clausewitzian terms, Napoleon lacked a clear culminating point of victory and staked everything on a single decisive battle.[30] Although that strategy had succeeded in past campaigns, his ego and obstinacy prevented him from recognizing that the invasion of Russia was unlikely to succeed. As Clausewitz described this dilemma:

> When in 1812 Bonaparte advanced on Moscow the critical question was whether the capture of the capital...would induce Czar Alexander to make peace...If, however, peace was not made at Moscow, Bonaparte would have no choice but to turn back...no matter how much more successful the advance on Moscow might have been, it would still have been uncertain whether it could have frightened the Czar into suing for peace. And even if the retreat had not led to the annihilation of the army, it could never have been anything but a major strategic defeat.[31]

Failing to achieve a decisive result at Borodino, the Grand Army was destroyed by a combination of scorched-earth tactics, vast distances, hunger, desertion, and bitter cold.

Similarly, during the Hundred Days, Napoleon's judgment was erratic in ways that ultimately undermined his military genius. Clinging to any desperate chance for victory, he hoped to win an improbable series of decisive battles and once again reign as the warlord of Europe. In an almost sympathetic way, Clausewitz describes this final gamble as both detached from reality and yet rational, given Napoleon's position:

> How could Bonaparte be blamed for not avoiding a battle...that was the only way to attain his objective: to cling to his last hope, to try to hold on to fortune's weakest threads...Should he have let mere danger scare him into this certainty? No, there are situations which the greatest caution could only be found in the greatest boldness, and Bonaparte's was one of them.[32]

Although Napoleon won a tactical victory at the Battle of Quatre Bras on June 16, 1815, this success was short-lived; two days later, he suffered final defeat at Waterloo, and his dreams of renewed glory came to an end. Even if he had prevailed at Waterloo, he would still have faced massive armies advancing on France. To survive, he would have needed to defeat each in turn—a virtually impossible task. While it may be overly harsh to fault Napoleon for this last gamble, his military genius had clearly been compromised by his egotism and a lack of strategic restraint.

In highlighting the failures of Napoleon, Clausewitz was doing much more than relishing the downfall of his nemesis. He was offering a cautionary lesson to aspiring military geniuses: if even Napoleon could falter, so could they. The implicit warning was clear—approach war with humility, prudence, and sound judgment.

Can You Achieve Military Genius?

Few books have attracted more aspiring military geniuses than *On War*. Its study has become almost *de rigueur* for military officers and armchair strategists alike, making it perhaps inevitable that readers compare themselves to Clausewitz's model of genius. A closer reading, however, yields a sobering conclusion: Clausewitz believed that true military genius was exceedingly rare. Yet, he also maintained that, through hard work, dedication, and the study of theory—particularly the ideas in *On War*—individuals could refine and elevate their innate abilities.[33] Reflecting on this process of self-cultivation, Clausewitz wrote that a would-be genius "extracts the essence from the phenomena of life, as a bee sucks honey from a flower."[34]

The American general George S. Patton Jr. exemplifies an individual who combined exceptional natural ability with a lifelong commitment to study, reflection, and self-improvement in pursuit of greatness as a military leader.[35] Like Clausewitz's bee extracting nectar from a flower, Patton was an avid reader who amassed a personal library and eagerly absorbed its contents. While on his honeymoon to London, he purchased a copy of Clausewitz's *On War*.[36] Upon returning, he joined his cavalry unit for field exercises and brought the volume with him. Despite the physical demands of training, he demonstrated the intellectual discipline to begin reading the dense treatise. In a letter to his wife, Beatrice, he admitted, "Clausewitz is about as hard as any thing [*sic*] can well be and is as full of notes and equal abstruseness as a dog is of fleas."[37] Though he struggled with the translation, Patton became deeply engaged with the text and pursued a lifelong study of Clausewitz, acquiring multiple editions and annotating them extensively.

As a result of his intellectual efforts, Patton honed his considerable natural talents and became one of the most formidable commanders in American military history. More than any other Allied general during World War II, he demonstrated the *coup d'œil* to rapidly assess unfolding military situations, along with the determination and strength of mind to act decisively. Though Patton was deeply flawed in other respects, he clearly possessed a rare combination of practical experience, theoretical understanding, intuitive perception, and resolve that enabled him to attain military greatness.

No campaign better illustrates this cultivated genius than the Battle of the Bulge. Even before the German offensive began, Patton had anticipated a winter attack, correctly predicted that Bastogne would be a key target, and began formulating a counteroffensive plan. When the Germans launched their assault, he remained composed while other Allied commanders faltered, and used his force of personality and clarity of vision to accomplish what many deemed impossible. Patton's Third Army disengaged from its current operations, pivoted ninety degrees, marched more than one hundred miles through snow and ice, and launched an immediate attack into the flank of the German forces surrounding Bastogne. This remains one of the most impressive feats of American military leadership and a powerful example of how Patton's lifelong study and deliberate cultivation of military genius paid tangible dividends in combat.

While it may be unrealistic to expect everyone to become a Napoleon or a Patton, Clausewitz's conception of military genius offers encouragement to those seeking to cultivate their own innate talents. In this view, military genius is not wholly innate but can be developed—at least in part—through openness to learning, sound judgment, and experience, as much as through formal intellectual training.[38] As Clausewitz argued, art is trained ability.[39] Wisdom is available to those who seek it, and *On War* serves as a tool for developing one's own strategic capacity.

CONCLUSION: "SO WHY DO WE NEED CLAUSEWITZ?"

This book has sought to embrace complexity and context in order to offer multiple perspectives on Carl von Clausewitz's *On War*. It has not attempted to obscure Clausewitz's flaws or the limitations of his work. At the same time, it has engaged with critics who portray him as either an immoral warmonger (such as Liddell Hart and Keegan) or an outdated thinker whose ideas no longer apply to contemporary conflict (such as van Creveld and Kaldor). It has also addressed and corrected several common misinterpretations of his theories. Ultimately, this study aims to offer a balanced and accessible introduction to Clausewitz that serves as a foundation for further study and debate.

Ignore Clausewitz at Your Own Peril

This book began with a provocative question: Why do we need another book about Clausewitz? To answer this, it first confronted the many criticisms that have led some to question his continued relevance. It then traced how *On War* is frequently described as poorly written, incomplete, confusing, and—in some respects—an outdated artifact of Napoleonic warfare. If this critique holds, would it not be reasonable to consign Clausewitz to the margins of military thought?[40] Hopefully, by this point, the reader will answer "no"—or at least have a better-informed response to the question of whether Clausewitz's work remains valuable today.

To test the contemporary relevance of Clausewitz, let us consider a simple thought experiment, using the ongoing conflict between Ukraine and Russia. If Clausewitz's theories can meaningfully explain or anticipate key dynamics in this war, that would suggest his continued relevance. Conversely, if his ideas fail to account for central aspects of the conflict, then doubts about their applicability would be justified. Although it is still too early to know how this war will ultimately conclude, the first three years suggest that Clausewitz's theories possess considerable explanatory and predictive power.

To begin with, the Russian theory of victory appeared to rest on the assumption that massed forces could rapidly overwhelm Ukraine's outnumbered and ill-equipped defenders. This assumption proved incorrect for three reasons identified in *On War*—friction, the warning that wars are not won in a single blow, and the primacy of the defense. Unlike war on paper, friction disrupted the Russian campaign at nearly every turn—tires went flat, vehicles became mired in traffic jams of epic scale, and soldiers deserted. Russian plans for a rapid advance on Kyiv quickly unraveled as Ukraine fought back with unexpected tenacity, illustrating that wars cannot be won in a single blow. The Russians also experienced firsthand that offensive operations are inherently more difficult than defensive ones, as they struggled with extended supply lines, exposed flanks, and deteriorating morale.

Next, the Russians discovered that they could not neutralize Ukrainian centers of gravity. They failed to capture Kyiv, remove Volodymyr Zelensky from power, or break the Ukrainian people's will to resist. In strategic terms, they became stalled. More critically, Russian commanders ignored the reality that their offensive had reached its culminating point. Rather than consolidating or recalibrating, they launched a series of futile attacks that expended valuable resources while yielding little in return.

As Ukrainians rallied to defend their nation, they embodied the elements of Clausewitz's trinity—primordial violence, enmity, and passion—and demonstrated the enduring power of nationalism and people's war. Despite the brutality of the Russian assault, military force alone failed to achieve Russia's political objectives. Even the capacity for destruction eventually encountered its natural limits.

In the final analysis, Russia lacked a military leader capable of objectively assessing the situation—or of challenging Vladimir Putin's strategic miscalculations. As in Napoleonic France, the campaign was ultimately undermined by the ambitions and ego of a single dominant figure.

While this comparison necessarily oversimplifies both the Russia-Ukraine War and Clausewitz's theories, it nonetheless illustrates the enduring analytical richness of *On War*. The first three years of the conflict, in fact, seem to vindicate Clausewitz's continued relevance. Even if one views this as a *post hoc* rationalization that may overstate his explanatory power, it is still worth asking: could any other classic work of strategy—*The Art of War*, *The Peloponnesian War*, and the like —have done a better job of predicting or explaining the conflict? The answer is likely no.

If *On War* still holds analytical value, the question becomes how best to interpret and apply its insights today.

Embrace Context, Imperfection, and Complexity

Appreciating the value of *On War* requires more than just reading it—
it requires understanding the man behind it, accepting its flaws, and
engaging with its complexity.[41] Understanding Clausewitz is a journey,
not a destination. With that in mind, here are some helpful reminders
for that journey.

Context matters. Clausewitz wrote in a very specific historical moment
and was driven by a particular set of motivations. While it is not necessary
to remember every detail of his biography, a few key points help foster a
more sympathetic understanding of his work. He aspired to be a scholar
but became a soldier out of necessity. His intellectual disposition marked
him as something of an outsider in the Prussian officer corps, even as he
proved himself repeatedly in battle. He fought in dozens of engagements,
served with distinction, and developed a uniquely informed perspective
on the Napoleonic Wars. These experiences were directly incorporated
into his theoretical writings. His service with the Russian army also
earned him lasting suspicion in Prussia, reinforcing his outsider status.
On War itself was written in part out of frustration and as an attempt to
reclaim intellectual authority after feeling wronged and marginalized.
In sum, *On War* is deeply shaped by Clausewitz's lived experiences and
personal struggles. Keeping this in mind is essential to engaging with
his theories in a thoughtful and historically grounded way.

Acknowledge the imperfections of this work.[42] As emphasized throughout
this book, *On War* was left incomplete and published posthumously.
As such, readers must approach the text with an awareness of its imperfections and a willingness to make allowances for its unfinished state.
While it may be tempting to speculate about how Clausewitz might have
revised or clarified his arguments had he lived to complete the work,
readers should resist the urge to finish the text on his behalf.

Embrace the complexity of the text. Clausewitz's theories are inherently
complex, and he often expressed them in dense, elliptical prose. His
subjects—war, politics, and the nature of human conflict—resist simpli-

fication, and his use of dialectical reasoning can be difficult to follow. While simple answers may offer greater comfort, they are rarely accurate. Instead, readers should strive to engage seriously with the nuance and ambiguity that define *On War*, recognizing that its richness lies precisely in its refusal to offer easy solutions.

Read—and reread—the text. Clausewitz's theories are complex, and his writing style can be opaque. As a result, *On War* does not lend itself to rapid consumption. Instead, it demands careful, repeated reading and sustained reflection. This deliberate approach, while time-consuming, yields valuable insights and deepens one's understanding over time.

Avoid reading isolated quotations. Readers might be daunted by the dialectical nature of Clausewitz's works; nevertheless, they should avoid the temptation to read only short passages. It is easy to be misled by the fact that Clausewitz often presents extreme or ideal-type arguments in one passage, only to contradict them a few sentences later. Readers should resist the urge to jump to conclusions. Instead, they should engage with longer sections—or, ideally, the entire work—to avoid simplistic or misleading interpretations of Clausewitz's actual arguments.

Recognize the rarity of military genius. While Clausewitz's works have attracted more than their share of aspiring strategists, it is important to remember that he believed true genius to be exceptionally rare. This does not mean that one cannot improve one's abilities, but it does suggest that most readers do not possess the innate gifts of a Napoleon. A humble, self-aware approach is essential when studying and applying Clausewitz's theories.

Understand that On War is not a how-to guide. *On War* is a philosophy of war, not a prescriptive checklist. It can help readers better understand problems, but it cannot offer direct instructions for what to do.[43] Rather than prescribing answers, Clausewitz encourages readers to embrace complexity and context and to exercise independent judgment. This model is more difficult to apply—but also more realistic and adaptable.

Embrace the Challenge. Meaningful intellectual achievement demands sustained effort—and *On War* is no exception. Clausewitz's dense and occasionally contradictory prose has challenged generations of strategic thinkers. Yet that very difficulty invites serious readers to sharpen their thinking and deepen their understanding of strategy. Embracing the challenge is part of becoming a more thoughtful and capable strategist.

Restoring Clausewitz to Engage *On War*

Despite its many flaws, *On War* remains one of most profound works ever written on the nature of conflict. Unfinished, translated, and rooted in a specific historical moment, it nonetheless offers a deep and enduring framework for thinking about strategy, war, and politics. This book has sought to illuminate Clausewitz's ideas, contextualize his arguments, and guide readers through the complexity of his work. Far from consigning Clausewitz to the past, it has aimed to recover the vitality of his ideas, reestablish his place in today's strategic discourse, and serve as a critical companion for readers engaging with *On War* in all its depth and difficulty. May it inspire continued study and reflection—for Clausewitz's insights remain as essential now as ever.

Restoring Clausewitz remains a difficult but fulfilling task, for it is ultimately a call to serious thought—a reminder that the study of war demands both intellectual humility and enduring commitment.

Notes

1. For a discussion of why Clausewitz's concept of genius is critical to understanding *On War*, see Paret, "The Genius of *On War*" in Clausewitz, *On War* (Howard and Paret), 3–25.
2. Clausewitz, *On War* (Howard and Paret), III.17, 220.
3. Clausewitz, 140. For a theoretical discussion of the significance of leaders who go beyond traditional constraints in their decision-making, see Daniel and Smith, "Statesmanship and the Problem of Theoretical Generalization," 156–184.
4. Clausewitz, *On War* (Howard and Paret), VIII.2, 580; Clausewitz, *On War* (Howard and Paret), VII.22, 570; and Clausewitz, *On War* (Howard and Paret), VIII.3, 583.
5. Aron, *Clausewitz: Philosopher of War*, 134, Echevarria II, *Clausewitz and Contemporary War*, 16, and Ferguson, *Civilization*, 158.
6. Bassford, *Clausewitz in English*, 31
7. Paret, *The Cognitive Challenge of War*, 131.
8. Clausewitz, *On War* (Howard and Paret), I.3, 100–102.
9. Clausewitz 117. See also Echevarria, *Clausewitz and Contemporary War*, 102,109–114, and 118; Heuser, *Reading Clausewitz*, 72; Paret, *Clausewitz and the State*, 161; and Strachan, *Clausewitz's On War*, 95 and 127.
10. Boyd, *The Essence of Winning and Losing*.
11. Chandler, *The Campaigns of Napoleon*, and Duggan, *Napoleon's Glance: The Secret of Strategy*.
12. Clausewitz, *On War* (Howard and Paret), I.3, 100–112; Clausewitz, *On War* (Howard and Paret), VIII.1, 578; Clausewitz, *On War* (Howard and Paret), VIII.6, 606; and Clausewitz, *On War* (Howard and Paret), VIII.9, 634. See also Sumida, *Decoding Clausewitz*, 17, 130–131, 170. Interestingly, this concept was not original to Clausewitz; it had been employed by Frederick the Great in his 1747 *Instructions to his Generals*. Clausewitz may have adopted this phrasing to align his work with that of the revered Prussian monarch and make it more acceptable to his audience. Frederick the Great, *Instructions to His Generals*, 49.
13. Clausewitz, *On War* (Howard and Paret), I.3, 102.
14. Clausewitz, 108.
15. Clausewitz, 213.
16. Clausewitz, 517. See also Lebow, *The Tragic Vision of Politics*, 180.

17. Clausewitz, *On War* (Howard and Paret), III.5, 185. See also Sumida, *Decoding Clausewitz*, 117.

18. See generally: Muth, *Command Culture*.

19. Sumida, *Decoding Clausewitz*, 117.

20. Claxton, *Hare Brain, Tortoise Mind*, and Kahneman, *Thinking Fast and Slow*.

21. Klein, *The Power of Intuition*. The United States Air Force also discovered similar findings as part of its three-part *Project Red Baron* report on air-to-air combat during the Vietnam War. Watts, *Clausewitzian Friction and Future War*, 105n215.

22. Emphasis in original; Clausewitz, *On War* (Howard and Paret), VI.8, 389.

23. Clausewitz, 578.

24. Clausewitz, 102–103.

25. Clausewitz, 103.

26. Esdaile, *Napoleon's Wars*, 13 and 109 and Waldman, *War, Clausewitz, and the Trinity*, 79.

27. Clausewitz, *On War* (Howard and Paret), I.3, 111.

28. Strachan, *Clausewitz's On War*, 153.

29. Parkinson, *Clausewitz: A Biography*, 157.

30. Much of *On War*, Book VIII, is examining the conditions that enabled Napoleon to pursue decisive battle—and how, in the end, this strategy proved unsuccessful. Echevarria, *Clausewitz and Contemporary War*, 135.

31. Clausewitz, *On War* (Howard and Paret), II.5, 166.

32. Clausewitz, *On Wellington*, 159–160.

33. For a modern attempt to reconcile Clausewitz's theories with contemporary understandings of experts, see Killion, "Clausewitz and Military Genius," 97–100.

34. Clausewitz, *On War* (Howard and Paret), II.2, 146; and Strachan, *Clausewitz's On War*, 126.

35. Daniel, *Patton*, esp. 10–12, and 30; Daniel, *21st Century Patton*; and Nye, *The Patton Mind*.

36. Ironically, he may have misunderstood the Prussian theorist's argument for the restraint of violence, as evidenced by his marginal note: "bunk—always go to the limits." Whether this misreading stemmed from Patton's own views on war or from the limitations of the English translation remains unclear. What is clear, however, is clear that Patton dedicated significant time to the study of warfare in general and Clausewitz's theories in particular. Daniel, *Patton*, 12.

37. Daniel, 11.

38. Aron, *Clausewitz: Philosopher of War*, 134–135.
39. Clausewitz, *On War* (Howard and Paret), II.2, 148.
40. Leonard, "That Clausewitz-is-Irrelevant 'Hot Take,'" and Leonard, "You Really Think I'm Irrelevant? LOL."
41. Jervis, *System Effects*.
42. For a powerful counterargument that Clausewitz's work is so incomplete and internally inconsistent as to render it of little practical value for military professionals, see Fleming, "Can Clausewitz Save Us From Future Mistakes?" 62–77.
43. Bassford, *Clausewitz in English*, 111.

Bibliography

Alterman, Eric. "The Uses and Abuses of Clausewitz." *Parameters* 17, no. 2 (Summer 1987): 18–32.

Aron, Raymond. *Clausewitz: Philosopher of War*. Translated by Christine Booker and Norman Stone. Prentice-Hall, Inc., 1985.

Baldwin, Peter M. "Clausewitz in Nazi Germany." *Journal of Contemporary History* 16 (1981): 5–26.

Bassford, Christopher. *Clausewitz in English: The Reception of Clausewitz in Britain and America, 1815–1945*. Oxford University Press, 1994.

Bassford, Christopher. "John Keegan and the Grand Tradition of Trashing Clausewitz." *War in History* 1, no. 3 (November 1994): 319–336.

Beiser, Frederick C. *Enlightenment, Revolution, & Romanticism: The Genesis of Modern German Political Thought, 1790–1800*. Harvard University Press, 1992.

Bell, David A. *The First Total War: Napoleons Europe and the Birth of Warfare as We Know It*. Houghton Mifflin, 2007.

Bellinger, Vanya Eftimova. *Marie von Clausewitz: The Woman Behind the Making of On War*. Oxford University Press, 2016.

Bellinger, Vanya Eftimova. "The Other Clausewitz: Findings from the Newly Discovered Correspondence between Marie and Carl von Clausewitz" *Journal of Military History* 79, no. 2 (April 2015): 345–367.

Bellinger, Vanya Eftimova. "A Timid Staff Officer? Reassessing Carl von Clausewitz's Role in the Battles of Ligny and Wavre (June 16–20, 1815)." In *The Sword and the Spirit: Proceedings from War and Peace in the Times of Napoleon*, edited by Zack White, 83–97. Helion Publishing, 2021.

Bernhardi, Friedrich von. *Germany and the Next War*. Translated by Allen H. Powles. Longmans, Green, & Co., 1914.

Beyerchen, Alan D. "Clausewitz, Nonlinearity, and the Unpredictability of War." *International Security* 17, no. 3 (Winter 1992–1993): 59–90.

Blainey, Geoffrey. *The Causes of War.* 3rd ed. Free Press, 1988.

Bousquet, Antoine. *The Scientific Way of Warfare: Order and Chaos on the Battlefields of Modernity* Columbia University Press, 2009.

Bowden, Mark. *Black Hawk Down: A Story of Modern War.* Atlantic Monthly Press, 1999.

Boyd, John. *The Essence of Winning and Losing.* June 28, 1995. https://www.dnipogo.org/boyd/pdf/essence_of_winning_losing.pdf.

Brigety II, Reuben E. *Ethics, Technology, and the American Way of War: Cruise Missiles and US Security Policy.* Routledge, 2007.

Brodie, Bernard, ed. *The Absolute Weapon: Atomic Power and World Order.* Harcourt Brace and Company, 1946.

Brodie, Bernard. *War and Politics.* MacMillan, 1973.

Brose, Eric Dorn. *The Kaiser's Army: The Politics of Military Technology in Germany During the Machine Age, 1870–1918.* Oxford University Press, 2004.

Brown, Kent Masterson. *Meade at Gettysburg: A Study in Command.* University of North Carolina Press, 2021.

Chandler, David G. *The Campaigns of Napoleon.* Scribner, 1966.

Chickering, Roger. *Imperial Germany and the Great War, 1914–1918.* 3rd ed. Cambridge University Press, 2014.

Churchill, Winston. "Finest Hour Speech." June 18, 1940. *International Churchill Society.* https://winstonchurchill.org/resources/speeches/1940-the-finest-hour/their-finest-hour/.

Cimbala, Stephen J. *Clausewitz and Escalation: Classical Perspective on Nuclear Strategy.* Frank Cass, 1991.

Citino, Robert M. *The German Way of War: From the Thirty Years War to the Third Reich.* University Press of Kansas, 2005.

Clark, Alan. *The Donkeys: A Controversial Account of the Leaders of the British Expeditionary Force in France 1915.* William H. Morrow, 1961.

Clark, Christopher. *Iron Kingdom: The Rise and Downfall of Prussia, 1600–1947.* Belknap Press, 2006.

Clausewitz, Carl von. *Carl von Clausewitz: Historical and Political Writings.* Edited and translated by Peter Paret and Daniel Moran. Princeton University Press, 1992.

Clausewitz, Carl von. *Clausewitz On Small War.* Edited and translated by Christopher Daase and James W. Davis. Oxford University Press, 2015.

Clausewitz, Carl von. *On War.* Edited and translated by Michael Howard and Peter Paret. Princeton University Press, 1976.

Clausewitz, Carl von. *On War.* Translated by J. J. Graham. Edited by Anatol Rapoport. Penguin, 1968.

Clausewitz, Carl von. *On Wellington: A Critique of Waterloo.* Translated by Peter Hofschörer. University of Oklahoma Press, 2012.

Claxton, Guy. *Hare Brain, Tortoise Mind: How Intelligence Increases When You Think Less.* Harper, 2000.

Clodfelter, Mark. *The Limits of Airpower: The American Bombing of North Vietnam.* University of Nebraska Press, 1989.

Cohen, Eliot, and John Gooch. *Military Misfortunes: The Anatomy of Failure in War.* Free Press, 1990.

Coker, Christopher. *Barbarous Philosophers: Reflections on the Nature of War from Heraclitus to Heisenberg.* Columbia University Press, 2010.

Coker, Christopher. *Rebooting Clausewitz: On War in the 21st Century.* Oxford University Press, 2017.

Cormier, Youri. "Hegel and Clausewitz: Convergence on Method, Divergence on Ethics." *The International History Review* 36, no. 3 (2014): 419–44.

Craig, Gordon A. *The Germans.* Meridian, 1983.

Craig, Gordon A. *The Politics of the Prussian Army, 1640–1945.* Oxford University Press, 1956.

Daniel III, J. Furman. *21st Century Patton: Strategic Insights for the Modern Era.* Naval Institute Press, 2016.

Daniel III, J. Furman. *Patton: Battling with History.* University of Missouri Press, 2020.

Daniel III, J. Furman, and Brian A. Smith. "Burke and Clausewitz on the Limitation of War." *Journal of International Political Theory* 11, no. 3 (2015): 313–330.

Daniel III, J. Furman, and Brian A. Smith. "Statesmanship and the Problem of Theoretical Generalization." *Polity* 42, no. 2 (2010): 156–184.

Daniel III, J. Furman, and Paul Musgrave. "Synthetic Experiences: How Popular Culture Matters for Images of International Relations." *International Studies Quarterly*, 61, no.3 (2017): 503–516.

Dimitriu, George. "Clausewitz and the Politics of War: A Contemporary Theory." *Journal of Strategic Studies* 43, no. 5 (2020): 645–685.

Donald, David Herbert. *Lincoln.* Simon and Schuster, 1995.

Duggan, Willian. *Napoleon's Glance: The Secret of Strategy.* Nation Books, 2002.

Dunnigan, James F. *How to Make War: A Comprehensive Guide to Modern Warfare in the Twenty-First Century.* 4th ed. Quill, 2003.

Dupuy, T. N. *A Genius for War: The German Army and General Staff, 1807–1945.* Prentice-Hall, 1977.

Dylan, Bob. *Chronicles: Volume One.* New York: Simon and Schuster, 2004.

Echevarria II, Antulio J. *After Clausewitz: German Military Thinkers Before the Great War.* University Press of Kansas, 2000.

Echevarria II, Antulio J. *Clausewitz and Contemporary War.* Oxford University Press, 2007.

Echevarria II, Antulio J. "Clausewitz's Center of Gravity: It's Not What We Thought." *Naval War College Review* 56, no. 1 (Winter 2003): 108–123.

Engberg-Pedersen, Anders. *Empire of Chance: The Napoleonic Wars and the Disorder of Things.* Harvard University Press, 2015.

Esdaile, Charles. *Napoleon's Wars: An International History, 1803–1815.* Penguin, 2008.

Esdaile, Charles. *The Peninsular War: A New History.* St. Martin's Press, 2003.

Ferguson, Niall. *Civilization: The West and the Rest.* Penguin, 2011.

Findley, Timothy. *The Wars.* Delacorte Press, 1977.

Fleming, Bruce, "Can Clausewitz Save Us From Future Mistakes?" *Parameters*, 34, no. 1 (Spring 2004): 62–77.

Fleming, Colin M. *Clausewitz's Timeless Trinity: A Framework for Modern War.* Ashgate Publishing, 2013.

Fleming, Ian. *Moonraker.* MacMillan, 1955.

Foley, Robert T. *German Strategy and the Path to Verdun: Eric von Falkenhayn and the Development of Attrition 1870–1916.* Cambridge University Press, 2005.

Forester, C. S. *The Commodore.* Michael Joseph, 1945.

Frederick the Great. *Instructions to His Generals.* Translated by Thomas Phillips. Stackpole, 1960.

Fritz, Stephen. *The First Soldier: Hitler as Military Leader.* Yale University Press, 2018.

Fukuyama, Francis. *The End of History and the Last Man.* Free Press, 1992.

Fussell, Paul. *The Great War in Modern Memory.* Oxford University Press, 1975.

Gat, Azar. *A History of Military Thought: From the Enlightenment to the Cold War.* Oxford University Press, 2001.

George, Alexander L., and Andrew Bennett. *Case Studies and Theory Development in Social Sciences* MIT Press, 2005.

Goerlitz, Walter. *History of the German General Staff, 1657–1945.* Translated by Brian Battershaw. Praeger Press, 1953.

Goethe, Johann Wolfgang von. *Miscellaneous Travels of J. W. Goethe: Letters from Switzerland; the Campaign in France, 1792; the Siege of Mainz; and a Tour of the Rhine, Maine, and Neckar. 1814–15.* Edited and translated by L. Dora Schmitz. George Bell and Sons, 1884.

Griffith, Paddy. *Battle Tactics of the Civil War.* Yale University Press, 1989.

Hagemann, Karen. *Revisiting Prussia's Wars Against Napoleon: History, Culture and Memory.* Cambridge University Press, 2015.

Handel, Michael I., ed. *Clausewitz and Modern Strategy.* Frank Cass, 1986.

Handel, Michael I. *Masters of War.* 3rd ed. Frank Cass, 2001.

Hemry, John G. *Stark's Command.* Ace Books, 2001.

Hemry, John G. *Stark's Crusade.* Ace Books, 2002.

Hemry, John G. *Stark's War.* Ace Books, 2000.

Herberg-Rothe, Andreas. *Clausewitz's Puzzle: The Political Theory of War.* Oxford University Press, 2007.

Heuser, Beatrice. *Reading Clausewitz.* Pimlico, 2002.

Hirschman, Albert O. *The Passions and the Interests: Political Arguments for Capitalism before Its Triumph.* Princeton University Press, 1977.

Holmes, Andrew. *Carl Von Clausewitz's On War: A Modern-Day Interpretation of a Strategy Classic.* Infinite Ideas Limited, 2010.

Holsti, Kalev J. *War, the State, and the State of War.* Cambridge University Press, 1996.

Howard, Michael. *Clausewitz.* Oxford University Press, 1983.

Howard, Michael. *The Invention of Peace: Reflections on War and International Order.* Yale University Press, 2000.

Huber, Jeff. "Clausewitz is Dead." *Proceedings* 127, no. 3 (March 2001): 119–121.

Huntington, Samuel P. *The Soldier and the State: The Theory and Politics of Civil Military Relations.* Belknap Press, 1957.

Hurt, James. "Sandburg's Lincoln Within History." *Journal of the Abraham Lincoln Association* 20, no. 1 (Winter 1999): 55–65.

Jervis, Robert. *System Effects: Complexity in Political and Social Life.* Princeton University Press, 1997.

Jomini, Antonie Henri. *The Art of War: A New Edition with Appendices and Maps.* Translated by G. H. Mendell and W. P. Craighill. J. B. Lippincott & Company, 1862.

Kahn, David. "Clausewitz and Intelligence" in *Clausewitz and Modern Strategy*, edited by Michael I. Handel, 117–126. Frank Cass, 1986.

Kahn, Herman. *Thinking About the Unthinkable*. Horizon Press, 1962.

Kahneman, Daniel. *Thinking Fast and Slow*. Farrar, Straus, and Giroux, 2011.

Kaldor, Mary. *New & Old War: Organized Violence in a Global Era*. Stanford University Press, 1999.

Kaplan, Fred. *Daydream Believers: How a Few Grand Ideas Wrecked American Power*. John Wiley & Sons, Inc., 2008.

Keegan, John. *The Face of Battle: A Study of Agincourt, Waterloo, and the Somme*. Viking Press, 1976.

Keegan, John. *A History of Warfare*. Alfred Knopf, 1993.

Keegan, John. *The Price of Admiralty: The Evolution of Naval Warfare from Trafalgar to Midway*. Viking, 1989.

Killion, Thomas H. "Clausewitz and Military Genius." *Military Review* 75, no. 4 (1995): 97–100.

Kissinger, Henry A. *Nuclear Weapons and Foreign Policy*. Harper & Brothers, 1957.

Kitfield, James. *Prodigal Soldiers: How a Generation of Officers Born of Vietnam Revolutionized the American Style of War*. Brassey's, 1997.

Klein, Gary. *The Power of Intuition: How to Use Your Gut Feelings to Make Better Decisions at Work*. Random House, 2003.

Knight, Frank H. *Risk, Uncertainty, and Profit*. Houghton Mifflin Company, 1921.

Knox, MacGregor, and Murray, Williamson. *The Dynamics of Military Revolution, 1300–2050*. Cambridge University Press, 2001.

Kuhn, Thomas S. *The Structure of Scientific Revolutions*. 2nd ed., enlarged. University of Chicago Press, 1970.

Lebow, Richard Ned. *The Tragic Vision of Politics: Ethics, Interests, and Orders*. Cambridge University Press, 2003.

Lefebvre, Georges. *Napoleon: From 18 Brumaire to Tilsit, 1799–1807.* Translated by Henry F. Stockhold, Columbia University Press, 1990.

Leonard, Steve. "That Clausewitz-is-Irrelevant 'Hot Take' Isn't Blasphemous. It's Just Wrong." *Modern War Institute*, March 5, 2019. https://mwi.westpoint.edu/clausewitz-irrelevant-hot-take-isnt-blasphemous-just-wrong/.

Leonard, Steve. "'You Really Think I'm Irrelevant? LOL.' A Letter to Clausewitz Haters from Beyond the Grave." *Modern War Institute*, May 6, 2020. https://mwi.usma.edu/really-think-im-irrelevant-lol-letter-clausewitz-haters-beyond-grave/.

Levinger, Matthew. *Enlightened Nationalism: The Transformation of Prussian Political Culture, 1806–1848.* Oxford University Press, 2000.

Lewis, C. S. *That Hideous Strength: A Modern Fairy-Tale for Grown-Ups.* Scribner, 1945.

Liddell Hart, B. H. *The British Way of Warfare.* Faber & Faber, 1932.

Liddell Hart, B. H. *Foch: The Man of Orléans.* Greenwood Press Publishers, 1980.

Liddell Hart, B. H. *The German Generals Talk.* William Morrow and Company, 1948.

Liddell Hart, B. H. *Strategy: Second Revised Edition.* Meridian, 1991.

Lieven, Dominic. *Russia Against Napoleon: The True Story of the Campaigns of War and Peace.* Viking, 2010.

Luvaass, Jay. "Clausewitz, Fuller, and Liddell Hart." In *Clausewitz and Modern Strategy*, edited by Michael I. Handel, 197–212. Frank Cass, 1986.

Luvaass, Jay. "Student as Teacher: Clausewitz on Fredrick the Great and Napoleon." In *Clausewitz and Modern Strategy*, edited by Michael I. Handel, 150–170. Frank Cass, 1986.

Machiavelli, Nicolo. *Discourses on Livy.* Translated by Harvey C. Mansfield and Nathan Tarcov. University of Chicago Press, 1996.

Machiavelli, Nicolo. *The Prince.* 2nd ed. Translated by Harvey C. Mansfield. University of Chicago Press, 1998.

Mann, Steven. "Chaos Theory and Strategic Thought." *Parameters* 22, no. 1 (Autumn 1992): 54–68.

McPhee, Peter. *Liberty or Death: The French Revolution.* Yale University Press, 2016.

McPherson, James M. *Tried by War: Abraham Lincoln as Commander in Chief.* Penguin, 2008.

Mearsheimer, John J. *Liddell-Hart and the Weight of History.* Cornell University Press, 1988.

Metz, Steven. "A Wake for Clausewitz: Toward a Philosophy of 21st Century Warfare." *Parameters* 24. no. 1, (Winter 1994–95): 126–132.

Morgan, J. H. *Assize of Arms: Being the Story of the Disarmament of Germany and Her Rearmament, (1919–1939).* Methuen & Co., 1945.

Münkler, Herfried. *The New Wars.* Polity Press, 2005.

Muth, Jörg. *Command Culture: Officer Education in the U.S. Army and the German Armed Forces, 1901–1940, and the Consequences for World War II.* University of North Texas Press, 2011.

Nester, William R. *Napoleon and the Art of Diplomacy: How War and Hubris Determined the Rise and Fall of the French Empire.* Savas Beatie, 2012.

Novo, Andrew R. and Jay M. Parker. *Restoring Thucydides: Testing Familiar Lessons and Deriving New Ones.* Cambria Press, 2020.

Nye, Roger H. *The Patton Mind: The Professional Development of an Extraordinary Leader.* Avery Press, 1993.

Osgood, Robert E. *Limited War: The Challenge to American Strategy.* University of Chicago Press, 1957.

Owens, William. *Lifting the Fog of War.* Farrar, Straus, and Giroux, 2000.

Paley, Norton. *Clausewitz Talks Business: An Executives Guide to Thinking Like a Strategist.* CRC Press, 2015.

Paret, Peter. *Clausewitz and the State: The Man, His Theories, and His Times.* Princeton University Press, 1985.

Paret, Peter. *Clausewitz in His Time: Essays in the Cultural and Intellectual History of Thinking about War.* Berghahn, 2015.

Paret, Peter. "Clausewitz: 'Half against My Will, I Have Become a Professor.'" *Journal of Military History* 75, no. 2 (April 2011): 591–601.

Paret, Peter. *The Cognitive Challenge of War: Prussia 1806.* Princeton University Press, 2009.

Paret, Peter. "Translation, Literal or Accurate." *Journal of Military History* 78, no. 1 (January 2014): 1077–1080.

Paret, Peter. *Yorck and the Era of Prussian Reform, 1807–1815.* Princeton University Press, 1966.

Parker, Geoffrey. *Global Crisis: War, Climate Change and Catastrophe in the Seventeenth Century.* Yale University Press, 2014.

Parkinson, Roger. *Clausewitz: A Biography.* Stein and Day, 1970.

Pinker, Steven. *Sense of Style: A Thinking Person's Guide to Writing in the 21st Century.* Penguin, 2015.

Porch, Douglas. "Clausewitz and the French, 1871–1914." In *Clausewitz and Modern Strategy,* edited by Michael I. Handel, 287–302. Frank Cass, 1986.

Powell, Colin, and Joseph E. Persico. *My American Journey: An Autobiography.* Random House, 1995.

Reardon, Carol. *With a Sword in One Hand and Jomini in the Other: The Problem of Military Thought in the Civil War North.* University of North Carolina Press, 2012.

Rhodes, Richard. *The Making of the Atomic Bomb.* Simon and Schuster, 1986.

Robertson, William Glenn. *The Staff Ride.* The Center for Military History, 1987.

Roland, Charles P. *An American Iliad: the Story of the Civil War.* McGraw-Hill, 1991.

Sandburg, Carl. *Abraham Lincoln: The War Years, 4 Vol.,* Harcourt Brace, & Company, 1939.

Scheipers, Sibylle. *On Small War: Carl von Clausewitz and People's War.* Oxford University Press, 2018.

Schelling, Thomas C. *Arms and Influence.* Yale University Press, 1966.

Scott, Tony. *Crimson Tide*. Buena Vista Pictures Distribution, 1995.

Sherman, Nancy. *Stoic Warriors: The Ancient Philosophy Behind the Military Mind*. Oxford University Press, 2005.

Simpson, Emile. *War from the Ground Up: Twenty-First-Century Combat as Politics*. Oxford University Press, 2013.

Smith, Rupert. *The Utility of Force: The Art of War in the Modern World*. Vintage Books, 2008.

Snyder, Jack. *Ideology of the Offensive: Military Decision Making and the Disasters of 1914*. Cornell University Press, 1984.

Staudenmaier, William O. *Vietnam: Mao vs. Clausewitz*. US Army War College, 1976.

Stephenson, Nathanial Wright. *Lincoln: An Account of His Personal Life, Especially of its Springs of Action as Revealed and Deepened by the Ordeal of War*. Bobbs-Merrill Company,1922.

Stoker, Donald. *Clausewitz: His Life and Work*. Oxford University Press, 2014.

Strachan, Hew. *Clausewitz's On War: A Biography*. Atlantic Monthly Press, 2007.

Strachan, Hew. "Michael Howard and Clausewitz." *Journal of Strategic Studies* 45, no. 1 (2022): 143–160.

Sumida, Jon Tetsuro. "A Concordance of Selected Subjects in Carl von Clausewitz's *On War*." *The Journal of Military History* 78 (January 2014): 271–331.

Strachan, Hew. *Decoding Clausewitz: A New Approach to On War*. University Press of Kansas, 2008.

Summers, Harry G. "Introduction." In *War, Politics, and Power: Selections from On War and I Believe and Profess*, edited by Edward M. Collins. Regnery Publishing, 1997.

Summers, Harry G. *On Strategy: A Critical Analysis of the Vietnam War*. Presidio Press, 1982.

Summers, Harry G. *On Strategy II: A Critical Analysis of the Gulf War*. Dell Publishing, 1992.

United States Marine Corps. *FMFM 1: Warfighting*. Headquarters, U.S. Marine Corps, 1989. https://www.marines.mil/Portals/1/Publications/MCDP%201%20Warfighting.pdf

Tuchman, Barbara. *The Guns of August*. Random House, 1962.

Turner, Stansfield. "Address to Chicago Council Navy League of the United States." March 9, 1973. https://www.cia.gov/readingroom/docs/CIA-RDP80B01554R003500280001-7.pdf.

Van Creveld, Martin. *Command in War*. Harvard University Press, 1985.

Van Creveld, Martin. "The Eternal Clausewitz." In *Clausewitz and Modern Strategy*, edited by Michael I. Handel, 35–50. Frank Cass, 1986.

Van Creveld, Martin. *The Transformation of War: The Most Radical Reinterpretation of Armed Conflict Since Clausewitz*. Free Press, 1991.

Van Creveld, Martin. "The Transformation of War Revisited." *Small Wars and Insurgencies* 13 (Summer 2002): 3–15.

Vidal, Gore. *Lincoln: A Novel*. Random House, 1984.

Waldman, Thomas. *War, Clausewitz, and the Trinity*. Ashgate Publishing, 2013.

Wallace, Rodrick. *Carl von Clausewitz, The Fog of War, and the AI Revolution: The Real World is Not a Game of Go*. Springer International Publishing, 2018.

Wallach, Jehuda L. "Misperceptions of Clausewitz's *On War* by the German Military." In *Clausewitz and Modern Strategy*, edited by Michael I. Handel, 213–239. Frank Cass, 1986.

Waltz, Kenneth N. *Theory of International Politics*. McGraw Hill, 1979.

Watts, Barry D. *Clausewitzian Friction and Future War*. Revised ed. Institute for National Strategic Studies, 2004.

White, Charles Edward. *The Enlightened Soldier: Scharnhorst and the Militärische Gesellschaft in Berlin, 1801–1805*. Praeger, 1989.

Williams, Kenneth P. *Lincoln Finds a General: A Military Study of the Civil War*. 5 vols. Macmillan, 1949–1957.

Williams, T. Harry. *Lincoln and His Generals*. Alfred A. Knopf, 1952.

Willmott, H. P., and Michael B. Barrett. *Clausewitz Reconsidered.* Praeger Security International, 2010.

Index

www.ingramcontent.com/pod-product-compliance
Lightning Source LLC
Chambersburg PA
CBHW060254220326
41598CB00027B/4105